CW00348378

HOMO ROBOTICUS

COLOPHON

Original title: *Homo Roboticus. 30 vragen en antwoorden over mens, robot & artificiële intelligentie*

Translated by: Benedicte De Winter

Editing: An Jacobs, Lynn Tytgat, Michel Maus, Romain Meeusen, Bram Vanderborght

Authors: Katrien Beuls, Joost Brancart, Malaika Brengman, Thomas Crispeels, Sander De Bock, Laurens De Gauquier, Paul De Hert, Emma De Keersmaecker, Lars De Laet, Kevin De Pauw, Philippe De Sutter, Nico De Witte, Daniel De Wolf, Ann Dooms, Shirley Elprama, Katleen Gabriels, Jo Ghillebert, Marc Goldchstein, Luc Hens, Veerle Hermans, Rob Heyman, Mireille Hildebrandt, Jonathan Holslag, An Jacobs, Charlotte Jewell, Erika Joos, Dimokritos Kavadias, Eric Kerckhofs, Nina Lefeber, Dirk Lefeber, Johan Loeckx, Cathy Macharis, Joachim Mathieu, Michel Maus, Romain Meeusen, Marc Noppen, Ann Nowé, Jo Pierson, Lincy Pyl, Hubert Rahier, Werner Schirmer, Luc Steels, Eva Swinnen, Lynn Tytgat, Guy Van Assche, Jean Paul Van Bendegem, Greet Van de Perre, Stephanie van de Sanden, Jolien van Keulen, Jef Van Laer, Bram Vanderborght, Lieselot Vanhaverbeke, Christophe Vanroelen, Tom Verstraten, Lennert Vierendeels, Kim Willems

Cover design: Wouter De Raeve

Typesetting: www.frisco.be

© 2019 VUBPRESS
VUBPRESS is an imprint of ASP nv (Academic and Scientific Publishers)
Keizerslaan 34
1000 Brussels
Tel. +32 (0)2 289 26 56
Fax +32 (0)2 289 26 59
Email info@vubpress.be
www.vubpress.be

ISBN 978 90 5718 866 4
NUR 740, 980, 950
Legal deposit D/2019/11.161/019

No part of this publication may be reproduced and/or published by print, photocopy, microfilm, electronic or any other means without the prior written permission of the publisher.

HOMO ROBOTICUS

An Jacobs
Lynn Tytgat
Michel Maus
Romain Meeusen
Bram Vanderborght

CONTENTS

AUTHORS

Katrien Beuls
Joost Brancart
Malaika Brengman
Thomas Crispeels
Sander De Bock
Laurens De Gauquier
Paul De Hert
Emma De Keersmaecker
Lars De Laet
Kevin De Pauw
Philippe De Sutter
Nico De Witte
Daniel De Wolf
Ann Dooms
Shirley Elprama
Katleen Gabriels
Jo Ghillebert
Marc Goldchstein
Luc Hens
Veerle Hermans
Rob Heyman
Mireille Hildebrandt
Jonathan Holslag
An Jacobs
Charlotte Jewell
Erika Joos
Dimokritos Kavadias
Eric Kerckhofs
Nina Lefeber
Dirk Lefeber
Johan Loeckx
Cathy Macharis
Joachim Mathieu
Michel Maus
Romain Meeusen
Marc Noppen
Ann Nowé
Jo Pierson
Lincy Pyl
Hubert Rahier
Werner Schirmer
Luc Steels
Eva Swinnen
Lynn Tytgat
Guy Van Assche
Jean Paul Van Bendegem
Greet Van de Perre
Stephanie van de Sanden
Jolien van Keulen
Jef Van Laer
Bram Vanderborght
Lieselot Vanhaverbeke
Christophe Vanroelen
Tom Verstraten
Lennert Vierendeels
Kim Willems

ACKNOWLEDGEMENTS

Homo Roboticus is a call for greater collaboration and cooperation around technology. The book itself is an example of this kind of special partnership: more than fifty professors and researchers from the Vrije Universiteit Brussel (VUB) have shone their light on robotics and artificial intelligence from the perspective of their own very different disciplines, all based upon humanistic values that bind us together. We would like to thank all the authors who have described their individual visions and helped to create a unique collection of photographic materials.

The impetus for this book came from the VUB think-tank POINcaré, which was set up by rector Caroline Pauwels. Our thanks also go to her for taking part in the fight for a technological evolution centred around human beings. We also received support from the various vice-rectorates, and in particular from vice-rector of research Karin Vanderkerken and director of research Mieke Gijsemans. A great deal of support also came from BruBotics, the multidisciplinary research centre of the VUB, which specialises in human robotics, and from weKONEKT.brussels, a collaboration with the ULB that aims to bring together this discipline and the complex but fascinating reality of an international city like Brussels.

For us, the VUB is a wonderful and inspiring place to work. Whilst working on *Homo Roboticus* we received fantastic support from various departments such as the Audio Visual Services.

We would also like to say a heartfelt thank you to the various partners (Bloom Law, demens.nu, Innoviris and KultuurKaffee) and media partners such as Knack and the University of Flanders for setting a public debate in motion and drawing attention to this important subject. In particular, we would like to thank the publisher VUBPRESS for believing in our special story right from the start.

The original Dutch version of the book was proposed on 7 February in collaboration with the federal opera house De Munt. We owe particular thanks to Peter de Caluwe for opening up the stage to scientists, artists and policy makers. Special thanks also go to AI Prof. Dr. Luc Steels and Dr. Oscar Vilarroya for the performance of the opera Fausto which raises ethical questions about the future of AI and transhumanism.

This book is our attempt to reach out to everyone in our society. We will all feel the impact of robotics and artificial intelligence on our lives. Our thanks go out to the readers of this book who want to participate in this debate.

Using the two-armed YuMi® robot from 'You and Me', ABB allows human and robot to work together safely on the assembly of small products.

©Robotics ABB Benelux

FOREWORD

The POINcaré think-tank is a loosely bound collaboration between researchers with very different fields of daily work. What unites and drives them is that, inspired by the values that are specific to our university, they have an eye for what is at stake for our society today.

The teaching, research and service to society of the Vrije Universiteit Brussel are based on radical humanism. This translates into the principles of freedom, equality and solidarity, which run through the initiatives of the POINcaré think tank like a common thread.

Time and time again, we asked ourselves what the impact of expected developments would be on our values and how these values could best be safeguarded.

Following on from the success of the first project on the Human City comes this second initiative, *Homo Roboticus*.

Technology has the unique power to strengthen economic growth and transform societies, and this will be truer than ever as robotics permeates through our society. The applications are everywhere.

The advance of robotics offers hope, but also instils fear. Robots have the potential to massively boost our productivity, but they can also destroy jobs. They have applications that can make the lives of people with disabilities so much richer, but people can also become slaves to robots. There are self-driving vehicles, but also killer bots.

Technology has the ability to constantly improve and prolong human life, but also to make it truly unpleasant or even wipe it out altogether. The rise of robotics that we are currently seeing is no different. From a humanist perspective, our first priority must be to understand all the ins and outs of the subject and then to look for ways in which we can use this new technology in a humanistic way.

At the Vrije Universiteit Brussel, pioneering work has been done in the field of artificial intelligence and robotics. This work is characterised by its unique perspective: that of human-robot interaction based upon the question of how humans and robots can live side by side – naturally with the intention of increasing the potential for human development.

In this book, more than fifty professors and researchers from the Vrije Universiteit Brussel discuss the multidisciplinary facets of these developments. It has turned out to be a wonderful book that should appeal to everyone. It answers 30 provocative questions, and anyone who has read the book will undoubtedly be in a better position to form opinions on these subjects.

During the opening ceremony at the start of this academic year, I argued for the beauty, power and wisdom of science.

For the glory of wonder and imagination, of daring to think about and investigate subjects that no-one else was engaging with and of discovering how things work. For the power to translate these insights into solutions that work, that help us solve the problems of the world, and that increase our ability to control the forces of nature or use them for our own ends. And, finally, for the wisdom to be aware of your own power, of what you can and cannot do, of the consequences of your actions, and of what is desirable and what is not. The wisdom to retain an awareness of why you are doing what you are doing: to make society and the world a little better and, yes, a little more beautiful.

Homo Roboticus is a particularly fruitful illustration of that beauty, power and wisdom of science.

Caroline Pauwels,
Rector of the Vrije Universiteit Brussel

The original Dutch version of Homo Roboticus was proposed on 7 February in collaboration with the federal opera house De Munt. ©Jean Cosyn - VUB

INTRODUCTION

HOW CAN ROBOTS AND HUMANS LIVE SIDE BY SIDE?

The Ancient Greeks dreamed of them, Leonardo da Vinci made sketches of them, but it wasn't until 1921 that the word 'robot' was used for the first time. The term – which is derived from the Czech word for work or slave – was coined by Karel Čapek for his play *Rossum's Universal Robots*. It then took until 1961 before the first industrial robot, the Unimate, was installed in a car factory, where it was used for iron casting.

The field of robotics is thus more than half a century old and is mainly characterised by the caged robots that serve to automate production in factories. With 184 robots per 10,000 employees, which are mainly found in its automotive industry, Belgium is one of the top ten countries for robotics in the world. South Korea takes first place with 631 robots per 10,000 employees, and the world average is 69. Such robots perform tasks termed the three Ds: dangerous, dull and dirty.

The next generation of robots will increasingly find its way into all facets of our daily lives. In the 1980s, Bill Gates dreamed of a computer in every house, but we now have several per person and it is hard to imagine any sector without them. Will the same thing happen for robots? The McKinsey consultancy firm predicts that in 2025 the market for robots will be much larger than that for PCs today. Hopefully, many European and – why not? – Belgian companies will play a role in this robotic revolution, offering new products and services that bring with them jobs and prosperity. According to the *US Robotics Roadmap* produced by the leading American universities, the field of robotics is set to become as influential as the internet is today.

Looking back at history, technology has the unique power to strengthen economic growth and transform societies – you only need consider the impact that the car, the computer, the internet and the smartphone have had on our lives to see that this is true. Think of how technology has shifted the economic centre of gravity from agriculture to industry and then to services over the centuries. It is crucial that society, our economy and our political system are prepared for the new generation of technology.

As yet, we rarely encounter robots in our day-to-day lives. It is true that robotic vacuum cleaners and lawn mowers are lending a helping hand around the house and drones can be found in every toy shop, but these niche robots are merely a sign of the impressive possibilities that lie ahead.

Some believe they are showing us the way towards a robotic utopia, whilst others predict doom-laden scenarios in which robots steal our jobs or even take over the world. But it is we humans who invent and create technology and we have the power to shape our own future.

What is a robot?

So, what is a robot? It's a question that's not as easy to answer as you might think. For example, what's the difference between a robot and a computer, or between an automatic machine and a remote-controlled vehicle? A lot of robots, especially those we read about in science fiction, are the mechanical equivalents of people and animals. So, what is it that makes a person or animal unique? Our senses – eyes, ears, nose, tongue and skin – receive a huge amount of information about the outside world. The robotic equivalent is sensors such as cameras and microphones. Our senses are connected to our brains via nerves that resemble long wires. These transmit electrical signals to our brains, which then perceive these signals as stimuli. Only then do we become aware of what we are seeing and hearing, allowing us to think and decide what to do. A robot's brain is a computer, which runs a program that tells it how to respond to data from its sensors. Just as our brains make our muscles move, the robot's computer controls its motors. Our muscles allow us to navigate in and influence the world around us, and manipulate objects. Robots, on the other hand, use sensors to monitor their environment, a computer to think and motors to move around.

The sudden upsurge of robotics has been fuelled by progress in a whole range of technologies, and the field unites these various disciplines. Falling sensor sizes have allowed robots to perceive their environment much more accurately than before. For example, consider how many sensors there are in a smartphone: tiny cameras, microphones, GPS receivers, compasses, air pressure gauges, etc., all packed into a box less than one centimetre thick. The rapid processing of all this sensor data calls for a huge amount of computing power.

Honda's humanoid robot Asimo uses its sensors to recognise people, avoid obstacles and fetch coffee.

©Honda

A self-driving vehicle generates a gigabyte of data every second, which has to be processed in real-time. This information is used to decide how firmly the accelerator pedal needs to be pressed and what direction to steer in: instantaneous processing is essential because you can't afford to be driving blind for even a few seconds.

The joints in our bodies are guided by more than eight hundred muscles. Robots, too, need an electric motor or other drive system for each of their joints and, since a humanoid robot can easily have a few dozen of these, this calls for several powerful and energy-efficient motors. It is going to take a lot more research if we are to match the performance of biological muscles.

The choice of materials is also an important consideration for robots. New lightweight yet flexible materials are needed that allow robots to interact safely and gently with human beings. Materials are even being developed that can repair themselves, just as our bodies can heal wounds and fractures.

What is artificial intelligence?

The development of artificial intelligence has a powerful hold upon our imaginations. *Artificial intelligence*, or *AI*, is a collective term for techniques such as *machine learning*, *deep learning* and *neural networks* that aim to let computers learn from their experiences. Robotics and AI are often lumped together, but the combination of the two can open up incredible possibilities: for example, the cobot that can learn new skills from a human worker in a factory, the self-driving vehicle that learns from other cars, or the exoskeleton or portable robotic suit that learns to adapt itself to its human users.

However, not all robots necessarily incorporate AI in their coding. Classic industrial robots – for example, those that assemble parts in a factory – are fully pre-programmed and have no capacity to learn. Conversely, AI also exists independently of robots, where it may be used to calculate the best route from A to B, optimise search engines, recommend music and movies, or trade shares via computers.

Increasingly futuristic applications are also being developed, such as AI programs that advise doctors on patient diagnoses, sit on boards of directors or even write books. In this book we do not view these applications as robots, but we also know that the rise of robotics cannot be considered in isolation.

Moreover, robots and artificial intelligence are subject to a principle known as Moravec's paradox, whereby robots struggle with tasks that human beings find trivially easy but excel at things people find difficult. For example, the task of tying shoelaces, which most children can master with ease, is still incredibly challenging for a robot, yet a computer defeated the chess grandmaster Garry Kasparov back in the 1990s. It seems strange that walking, throwing a ball, picking up an object or handling a needle are so difficult for robots. One possible explanation is that these actions are underpinned by millions of years of evolution to improve and optimise the design of mankind, whilst skills such as mathematics and logic have evolved much more recently.

This all leads us to the conclusion that machines and human beings can actually make a great team. Their capabilities are highly complementary with people being handy, creative and versatile, whilst robots have the ability to process large amounts of information or tirelessly perform highly repetitive tasks and can be extremely precise and very strong. This is why research at the VUB focuses on applications based upon human-robot interaction in healthcare and manufacturing. Because human beings play a central role and are involved in the development process right from the outset, the role of the humanities and social sciences is extremely important.

With great power comes great responsibility

So, is it true that because man and robot work together so well we have nothing to worry about? That may be taking things a little too far. To quote Spiderman: *with great power comes great responsibility*. Robots are set to become a very powerful tool, which we must handle with care. As will be discussed in this book, robots will have an impact on our work, care system, domestic environment, leisure, education, security, entrepreneurship, law, taxation and society. Certain jobs will disappear, with occupations such as taxi or truck driver possibly set to go the way of the trades of farrier and candle maker in the past. But other jobs will simply change, as the secretaries of today perform different tasks than their predecessors did before the arrival of the PC. And new, as yet undreamt of professions will also spring up.

Because this technology is bound to have an impact upon our daily lives, we need to consider the ethics of these changes so that we can safeguard the rights of our citizens, such as privacy and security. But this should not lead to paralysis – instead we should try to take full advantage of the opportunities that arise, using them to tackle social challenges.

Freedom, equality and solidarity

More than 50 professors and researchers from the POINcaré think tank of the Vrije Universiteit Brussel (VUB) have gone in search of the Homo Roboticus. Based upon their own fields of research and interest, they describe their vision of who and what the Homo Roboticus is and how it can live in harmony with robots. The result is this multidisciplinary book on thirty thought-provoking questions. All these divergent visions are focussed on the basic values of radical humanism of the VUB and the city of Brussels: freedom, equality and solidarity.

For us, **freedom** means free research: the rejection of arguments that stem merely from authority and the guarantee of free judgement. But it also means the freedom of people with disabilities: by the use of an exoskeleton to compensate for paralysis, by giving an amputee a robotic foot or by allowing a blind person to use a car independently. It also stands for the freedom to live a life in which a robot fulfils our basic needs. As well as freedom, we also consider possible losses as a result of our growing dependence upon robotic solutions. Are we still free to do what we want if robots are constantly seeing and hearing what we do and don't do and violating our privacy?

For us, **equality** means equal status and the recognition of diversity. Will the assistance of pro bono robotic lawyers mean that justice is equally available to all? Will physical or mental differences be corrected by robotic assistants? Or will inequalities merely be exacerbated by the fact that some can afford robotic assistance and others cannot? And where will the prosperity created by robots end up? In the hands of a few, resulting in growing inequality? Or will society as a whole benefit?

Solidarity means our commitment to the major challenges facing society and our concern for the respectful treatment of our fellow human beings and the world as a whole. We are facing a range of challenges, such as the ageing population, rising health costs and the need for better and healthier jobs. Robots can be part of the answer but we also need to be alert to the ethical and safety issues that they raise.

UTOPIA / DYSTOPIA

The metal endoskeleton of the T-800 from *Terminator*.

© Usa-Pyon / Shutterstock.com

WHAT CAN SCIENCE FICTION CONTRIBUTE TO THE FIELD OF HUMANISTIC ROBOTICS?

By Prof. Dr. Dimokritos Kavadias and Jolien van Keulen

Even before robotics became an existing field of study, fiction had already predicted the existence of robots. Science fiction (SF), a genre that explores alternative worlds and futures, is rarely credited as being part of the 'canon' of literature. But enthusiasts see it as 'realistic speculation about possible future events, based solidly on an adequate knowledge of the real world, past and present, and on a thorough understanding of the nature and significance of the scientific method'. As technology advances and we move ever closer to the possibility of human-robot integration, authors and filmmakers have been exploring and depicting the advent of robots and its consequences for centuries: from non-human beings in Greek mythology to advanced artificial intelligence in Hollywood films.

Myths, sagas and, more recently, science fiction have been an outlet for speculation about what might be possible. But contrary to what the 'science' in SF might lead us to assume, such stories are not purely rational or speculative. They also express emotions. The stories in SF give us an insight into the way people see robots, the problems we envisage as a result of the advent of robots, and the fears that prevail – as well as a glimpse of more hopeful expectations. Understanding these human factors can be an important step in the process of robot design and the development of policy relating to robots. The question we are highlighting here is whether SF stories about robots can show us how to deal with this technology in a humanistic way.

As soon as we human beings give free reign to our imaginations, we start trying to shape the two basic emotions that dominate our lives: hope and fear. These two emotions repeatedly crop up as leitmotifs in stories about robots, which always seem to sketch out either a utopian or a dystopian world. Many fictional accounts of the mechanical simulation of life focus on the destructive consequences of robots. Man is characterized as 'good', while technology – represented by a robot – stands for evil. But SF also explores the more positive potential of robots.

'Hell to my left': the birth of the robot

The word 'robot' first appeared in 1920. In his play *R.U.R. – Rossum's Universal Robots*, Czech playwright Karel Čapek introduces autonomously thinking and acting instruments, designed for maximum work capacity. In old Czech, the word 'robota' means 'serf' or 'slave'.

> Robots of the world! We, the first union of Rossum's Universal Robots, declare man our enemy and outcasts in the universe.
> (...) You are ordered to exterminate the human race.
> Do not spare the men. Do not spare the women. Preserve only factories, railroads, machines, mines and raw materials.
> Destroy everything else. Then return to work.
> Work must not cease. – from R.U.R. Karel Čapek, 1920 – own translation.

As the above quote indicates, *R.U.R.* focuses on the destructive consequences of robots. This makes it a forerunner of the *Terminator* and *The Matrix* franchises, which show a future dominated by machines as the next step in evolution.

Just over a century before Čapek – in 1818 – the young Mary Shelley explored the same ideas in her pioneering book *Frankenstein or The Modern Prometheus.* Although *Frankenstein* is considered to belong to the horror genre, Shelley based her writing on the prevailing scientific insights of that time. In this classic novel, the young, brilliant doctor Victor Frankenstein succeeds in using scientific insights to simulate life.

His creation, or 'monster', destroys the arrogant Frankenstein and his family. The subtitle, *The Modern Prometheus*, was a vivid warning to hubristic doctors and engineers. That hubris is a key theme of both *Frankenstein* and *R.U.R.* Frankenstein sees himself as the creator of a new Adam, thinking – just like Prometheus – that he can do this with impunity, while Rossum (from *R.U.R.*) sees himself as a kind of scientific substitute for God. Disapproval of such arrogance is ingrained in the Classical cultural heritage. It can also be found in the Jewish and Christian cultural heritage: *thou shalt not take the place of God and create a thinking being.* If we do, there is a good chance that we will be digging our own graves.

Nowhere is this idea so aptly and fully expressed as in *Dune*, Frank Herbert's classic SF saga from the 1960s. The universe of *Dune* stands out in the SF genre because of the total absence of robots. This is because the world has been purged of thinking machines by a holy war (the *Butlerian Jihad*) based upon the commandment *'Thou shalt not make a machine in the likeness of a human mind'*. Although what is innovative about both *Frankenstein* and *R.U.R.* appears to be the description of autonomously acting creations, it is primarily the mental aspect – the autonomous thinking, the imitation of the human mind – that is perceived as threatening. It seems likely that these kinds of smart instruments will escape our control.

SF classics such as *Westworld* (a 1973 film and 2016 television series by HBO) or the superb *Blade Runner* (1982) do not directly address the fear of rebellion by robots, but they do deal with the possibility of robots accidentally or deliberately escaping our control. Since the 1980s, this fear has been translated into SF stories in which artificial intelligence breaks loose, heralding the end of humanity.

'Paradise to my right': The human in the robot (and the robot in the human)

One of the first explorations of the potential of robots as a source of hope can be found in the American science fiction of the post-war years. Isaac Asimov wrote his first robot story in 1939 at the age of nineteen, and continued to write on this subject until his death in 1992. His early stories act as a harbinger of the '*Golden Fifties*', a time when household appliances were being produced on a large scale. In these stories, robots are no longer soulless machines that want to dominate mankind, but the products of practical engineers. Robbie the babysitting robot feels very much like the ancestor of R2-D2: a caring and attentive side-kick for every child in the house.

That optimism of the 1950s is reflected in the robots of SF films such as *Star Wars*. The comic duo R2D2 and C-3PO (and, later on, BB8), as well as David, the robot that looks like a child in Steven Spielberg's *A.I.*, represent reliable allies in a turbulent world. They serve humans unconditionally, providing physical, material and emotional support.

SF may have the status of pulp literature, but the genre is also used to explore more philosophical questions about what it means to be human, and what differentiates people from machines. In robot stories that are not purely apocalyptic, the contradictions between man and robot are often elaborated, and the boundaries between the two explored. Themes such as emotions, creativity, rationality and irrationality, autonomy and mortality are addressed.

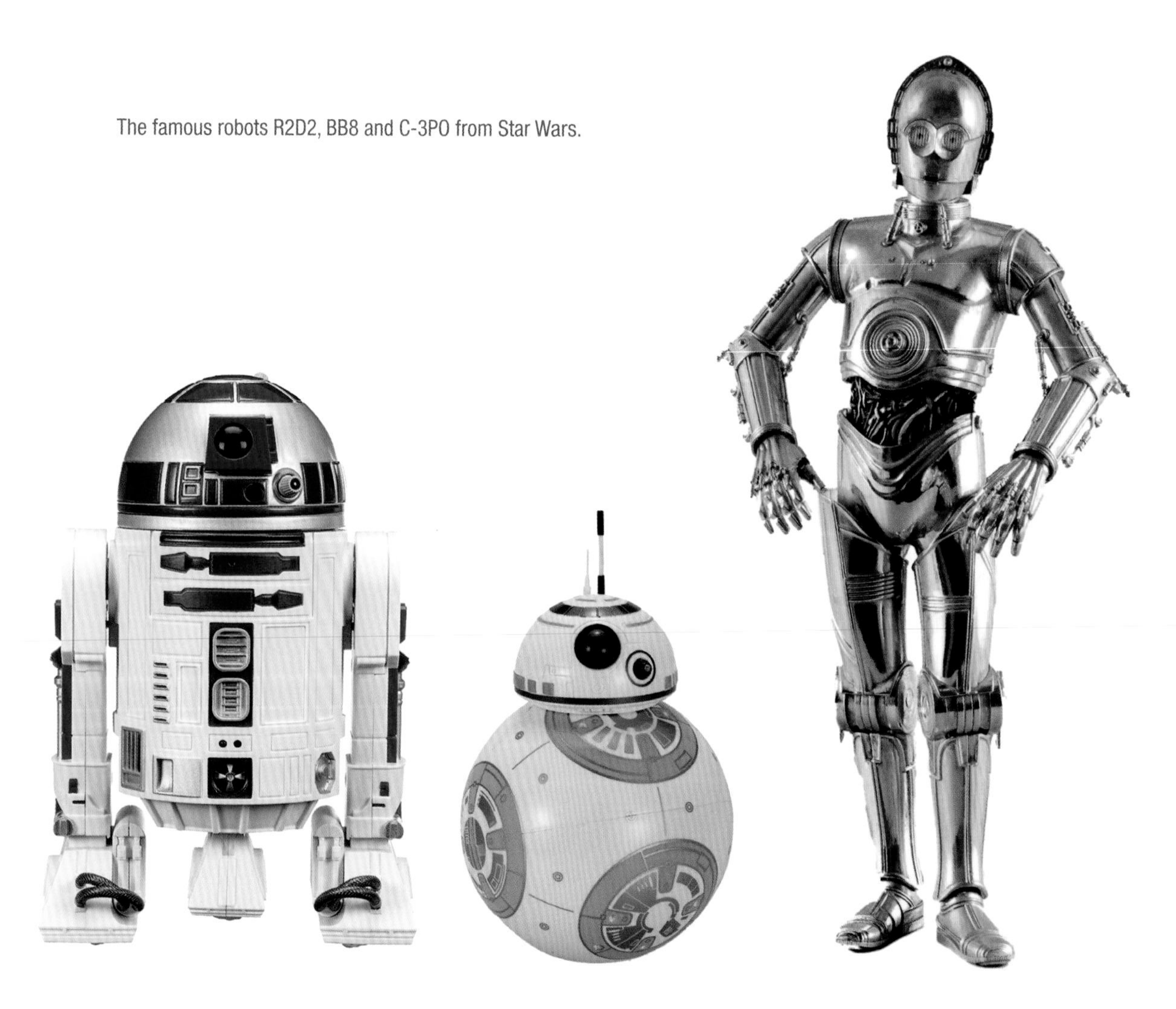

The famous robots R2D2, BB8 and C-3PO from Star Wars.

> And again he perceived himself *sub specie aeternitatis*, the form-destroyer called forth by what he heard and saw here. Perhaps the better she functions, the better a singer she is, the more I am needed. If the androids had remained substandard (...) there would be no problem and no need of my skill.
>
> — from *Do Androids Dream of Electric Sheep?* Philip K. Dick, 1968.

Frankenstein is an example of a story that explores the tension between robots and emotions. In the second part of the book (though not in popular film adaptations), 'the monster' largely speaks for itself. The creation experiences people's anger, sees poverty, learns to speak and write, and even succeeds in translating emotions into words. Shelley's story is thus more than simply a treatise against science. From the point of view of a thinking 'non-human', she describes happiness, sadness, fear and despair, and even pleads for humanity. The creature wants to escape loneliness and longs for a partner, a 'soul mate'. It is this demand for a second, female creation that faced Frankenstein with a dilemma. Should he continue helping a new species to grow, or opt to destroy his creation?

This quest for humanity in the robot can be seen in *R.U.R.* and, more recently, in the film *Blade Runner 2049*. In this film, the distinction between man and machine becomes razor-thin, since the reproduction of man and 'replicant' is seen as the salvation of mankind in the twenty-first century. In the context of ecological disaster, robots are the bringers of hope. Some stories also argue in favour of inalienable rights, by analogy with the extension of human rights to animal rights.

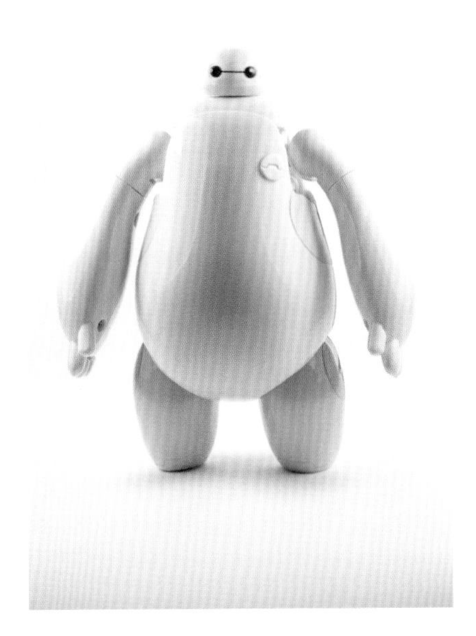

Baymax from the animated film Big Hero 6.
©Bruno Ismael Silva Alves / Shutterstock.com

Besides questions about what it means to be human, and the limits of human existence, robot stories regularly deal with the issue of whether humans are dependent on the robot, or vice versa. Who has power over whom? Robots were created to do the heavy work: in *R.U.R.* they are the only ones producing anything. But that also makes them indispensable. Do robots meticulously follow the programming that people have devised for them, or can they break free from it? Can their programming contain mistakes? Can they overrule people? Will they ever be able to 'hack' the human mind? How do we, as human beings, deal with 'instruments' that might be able to think for themselves? Do we destroy them completely, as in *Dune*, do we make ourselves superfluous as a species, or do we opt for the fusion of human and technology?

Towards humanistic robotics?

SF stories highlight the far-reaching consequences that scientific progress can have for individuals, societies, or even the whole of humanity. SF not only serves as entertainment, it also outlines the real, ethical dilemmas we face as a society. Since the Enlightenment, people have been drawn to the promise of progress. However, the fear of robots in SF highlights a fundamental dilemma: as human beings we want 'smart' instruments, machines that can take over our work both physically and mentally, and that can even think autonomously. But, at the same time, we fear these 'superior' AI creations since we actually want them to remain 'inferior' instruments. Whether we are aware of it or not, we are afraid that autonomous thinking will lead to full autonomy and a desire for freedom.

The American SF author Isaac Asimov attempted to analyse the possible dangers and benefits of robots in detail. In his stories he asks questions, such as: What does it mean to have a machine weigh up risks? What moral dilemmas should we address when we call upon machines to evaluate lives? Which criteria can produce optimal solutions for combating crime by robots? Can we trust machines to administer justice? Although Asimov is usually considered to be one of the 'hard SF' writers, he devotes a great deal of attention to the possible psychological and social consequences of robots. Robots – and the robot psychologist Suzan Calvin, who features in many of Asimov's robot stories – are thought constructs that allow us to reflect on what it means to be human, and on how we can successfully evolve as a species if we want to continue to integrate technology into our culture, up to – and including – the thorny issue of sex with robots.

Asimov describes robots as useful instruments that, due to their complexity and potential for social disruption, require a minimum number of moral rules. It is this attachment to humanity, to the project of a humanistic civilization, that led Asimov to formulate the so-called 'Laws of Robotics' in 1942:

1. A robot may not injure a human being or, through inaction, allow a human being to come to harm.
2. A robot must obey orders given it by human beings except where such orders would conflict with the First Law.
3. A robot must protect its own existence as long as such protection does not conflict with the First or Second Law.

It is this exploration of ideas that can make SF a genuine blueprint for the development of humanistic robotics. In one of his last robot books, *Robots and Empire*, Asimov introduces a so-called 'Zeroth Law': *A robot may not harm humanity, or, by inaction, allow humanity to come to harm.* If we, as scientists, explore these thoughts further, it can help us to apply the necessary wisdom as we push back the technical boundaries to the progress of mankind.

The robots in the stories of SF author Isaac Asimov are bound by the famous 'Laws of Robotics'. Each story describes how these laws play out in a particular situation.

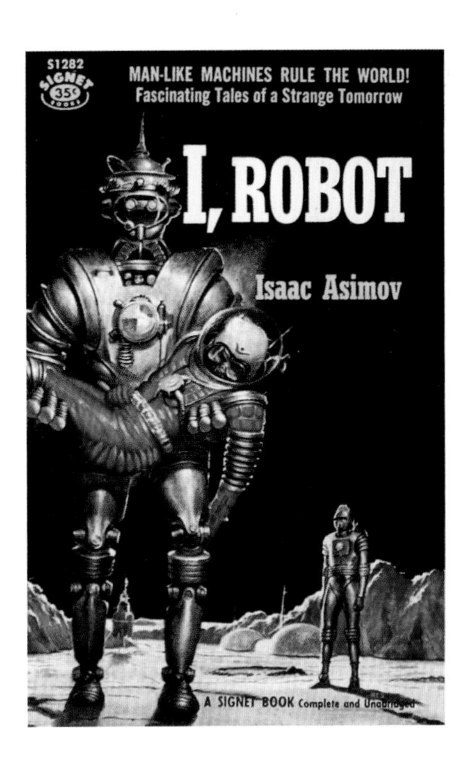

HUMANOID

In 2011, Robonaut 2 flew out to the International Space Station (ISS) to assist the astronauts in space.

©NASA

HUMAN OR ROBOT: WHICH MAKES THE BEST MACHINE?

By Dr. Ir. Tom Verstraten

In science fiction films, the scene is already set: robots will be stronger than people and better able to sense their environment; they will make smarter decisions and be equipped with all kinds of built-in technological gadgets. But how does this measure up to reality? Will robots ever reach the stage where we can barely distinguish them from people? Or will they even become superior to us humans, eventually overthrowing us?

Exceptional performance by robots and human beings

Have you ever gazed in awe at the spectacular leaps of a parkour artist, the graceful movements of a ballet dancer, or the breath-taking somersaults of a circus acrobat? A robot's movements, in contrast, are much less natural. Only the very latest robots can come anywhere close to imitating a human being.

Recently, Boston Dynamics shook the robotics community by introducing a robot that could perform a back flip, a feat that would be unthinkable for most other robots. Nevertheless, even the slightest change in the environment – a pebble, a light breeze, or a displaced mat – would have been enough to cause the stunt to fail.

Robots in factories are programmed to perform a well-defined movement with great precision. That's what they excel at. A human being can have a bad day where they work a bit more slowly than usual, or make the odd mistake. Industrial robots, on the other hand, can effortlessly assemble hundreds of cars in exactly the same way, at exactly the same speed, every time.

The problem is that robots, unlike humans, cannot adapt to changes in their environment. If the car shifts on the conveyor belt by just a few centimetres, the robot will try to screw on the car door in the wrong place. And if an unexpected obstacle appears in the path of the robotic arm, the robot will simply use all its strength to push it out of the way – with potentially disastrous consequences. That's why people and robots are kept strictly separated in factories.

©Darpa

Human environments, in particular, represent a major challenge for robots. For example, two-legged robots find it extremely difficult to walk on uneven terrain. Whereas we humans instantly sense when we are losing our balance and make the necessary corrections, robots struggle to assess the appropriate response. This became painfully clear during the Darpa Robotics Challenge of 2015, in which robots were required to perform simple tasks such as walking over uneven terrain, opening doors, and turning a knob. More than once, these high-tech robots – with a price tag of around a million dollars – went crashing to the ground.

Another difficult task is grasping objects. To do this, you not only have to shape your fingers to the object in question, you also need to be able to assess its firmness and slipperiness. If you don't squeeze your fingers hard enough, the object slips out of your hand. Squeeze too hard, and you're left with nothing but debris or pulp. Finding the right balance is something that human beings do intuitively, but which robots find particularly difficult.

In terms of pure power, however, the performance of robots is impressive. There are industrial robots that can lift loads of up to 2,300 kilograms – almost ten times more than the current world record for weightlifting held by the Russian Alexey Lovchev (264 kilos). But if you compare that with the weight of the robot itself, the performance suddenly seems a lot less impressive. A robot typically weighs five to ten times more than the maximum weight it can lift. For an average person, this ratio is about one to one.

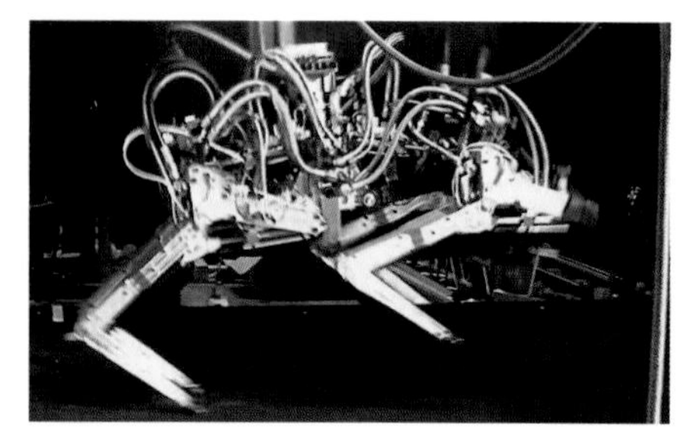

©Boston Dynamics

And what about speed? The fastest robot in the world is the Cheetah, once again from Boston Dynamics. This four-legged robot can reach speeds up to 45 km per hour. That's about as fast as the fastest man on Earth, Usain Bolt, but it's a long way off the performance of a real cheetah, which can achieve speeds of up to 100 km per hour. And four-legged robots are faster than two-legged ones: the top speed of Honda's Asimo is barely

9 km per hour. So human beings and animals can still outpace their robotic equivalents by some margin.

The examples described above show that people still outperform robots in many areas. The explanation for this lies in the unique properties of our natural 'motors': the muscles.

The search for the robotic muscle

Humans and animals move by contracting muscles in their bodies. These muscles are connected via tendons to the part of the body being moved – often this is a bone, but it could also be an eyelid or heart valve. When the muscle contracts, the tendon stretches, just like a spring or an elastic band. It is this elasticity that explains many of humankind's impressive achievements.

Elastic parts act as energy buffers: when stretched, they store energy, which they then release when they relax. Thanks to the elasticity of tendons, joints can absorb large amounts of energy over a short period of time. This is necessary to absorb shocks – for example, when landing after a high jump. Tendons can also release energy at appropriate times. This allows human beings to produce much more power than would be possible with their muscles alone, and is the reason for humans being able to produce extremely high forces in relation to their own weight, and outrun the fastest robots.

Humans and animals also use their muscles and tendons to adjust the stiffness of their joints. Contracting several muscle groups around a joint at the same time (co-contraction) can make the joint stiffer. If you do this with your elbow, for example, it makes it more difficult for another person or object to push your arm away. It also makes your movements more precise. Try pointing out a precise point on a map: you will notice that you try to keep your wrist and fingers stiff. By contrast, there are sometimes advantages to making your joints more compliant. When you jump from a height, for example, you make your knees more flexible by bending them slightly to help your body absorb the shock upon landing.

Besides their elasticity, there is another reason for the remarkable achievements of our muscles. They are made up of millions of muscle fibres – tiny motors that each contribute a small amount to the force exerted by the muscle. The more muscle fibres are active, the greater the force. Muscle fibres also specialize: some are better at fast movements but deliver little power, others deliver great power but are slower.

You can see which type of muscle fibre is which from their colour. Chickens (white meat) have many fast muscle fibres, while horses (red meat) have slow, powerful muscle fibres. By getting the two types of muscle fibre to work well together, you can achieve the exceptional performance we described earlier.

It may come as no surprise that roboticists draw inspiration from human muscles in their search for the ideal power train. For example, a great many robot drives already incorporate springs. It has been demonstrated that such drives perform particularly well in prostheses and exoskeletons, and in tasks such as throwing, carpentry or kicking a ball. The action of muscle fibres can also be imitated by having a large number of electric motors working together. Because concepts involving motors often end up being quite large and heavy, researchers are also experimenting with new power sources and materials. Some are even attempting to imitate muscle tissue.

Robots need senses too

Even if robots had the same muscles as human beings, we would still outperform them by a good margin in many areas. For example, we are also better at detecting and processing information from our environment.

Human beings have millions of tiny receptors in their bodies that enable them to sense their environment. The stimuli picked up by these receptors are transmitted to the brain via a complex network. This, too, poses a problem for robot designers. Although the artificial equivalent of human receptors does exist in the form of sensors, these are usually larger than their human counterparts. It is possible to produce tiny cameras and microphones, but the sensors responsible for touch, for example, are difficult to miniaturize. Unfortunately, these sensors are present in large numbers throughout the human body, making it difficult to integrate similar sensors into robots in comparable numbers. And even if it were possible, imagine how many cables would be needed to transmit the information from these sensors to the central processing unit – the robot's brain.

The solution is obvious: use fewer sensors. But that comes at the expense of reliability. If you have millions of receptors for touch, you can manage perfectly well if one fails. If you have only a few, it's a different matter. In addition, people are very good at adapting when one receptor – or an entire sense – is not working properly. But teaching a robot the difference between a defective camera and a dark room, or between dirt on a lens and a large obstacle in its way, is a lot more difficult. Just ask the makers of self-driving cars: it's one of their greatest challenges.

Human and artificial intelligence

Receiving information from thousands of sensors is one thing, processing it quite another. This brings us to the realm of artificial intelligence. Computer intelligence has come on in leaps and bounds in recent years. Nowadays, computers are able to carry out tasks that are far too complicated for most human beings. Solving a complex mathematical problem or winning a game of chess against a seasoned player? No problem. But when a robot is asked to differentiate between objects, recognize voices, have a conversation or tie laces, its performance suddenly drops below that expected of a five-year-old child.

Why are robots so good at complex tasks, when they mess up such seemingly simple ones? Tasks related to perception, motor skills, social skills and creativity are a piece of cake for humans – not because the tasks themselves are simple, but because evolution has been training us to perform them for thousands of years. Solving mathematical problems, on the other hand, requires abstract thinking. It is only over the past hundred years or so that the average human has been confronted with such tasks, so we find them much more difficult. For the computer, fully programmed as it is with mathematical algorithms, this is precisely the way of thinking that comes most naturally.

So what conclusion can we draw from all of this? You can't really say that people are smarter than computers or vice versa. We just think in a different way.

Robots and people: the perfect team?

Robots may never be able to do the same things as people, but maybe that's not what is needed. Human beings and robots complement each other. Robots are very good at repetitive tasks that bore humans to tears and they can carry loads far too heavy for the average person without ever tiring. Human beings, on the other hand, can perform precise and sophisticated tasks and use their creativity to engage with complex environments in which robots would run amok. Many jobs call for a combination of all these skills, so collaboration between robots and human beings is the way to go.

Ping pong for robots

A robot arm learns to play table tennis by reinforcement learning with the assistance of a human being.

©Axethe Max Planck Society

ARE ROBOTS SMARTER THAN PEOPLE?

By Prof. Dr. Ann Nowé

It finally happened on 15 March 2017. The eighteen-time Go world champion Lee Sedol was defeated by the AI system AlphaGo in a series of five games in Seoul. The final result of the tournament was 4-1 to AI. Go is a Chinese board game that has been played for more than 2,500 years. Given the complexity of the game, it was thought that this breakthrough would not be possible until the next decade. The number of possible configurations is greater than the number of atoms in our world, or, to put it another way, 10^{100} times greater than the number of configurations in a game of chess. This complexity explains why a solution based on pure computing power, as has been used for chess in the past, is not the answer in the case of Go. So how did Google DeepMind tackle the development of AlphaGo?

AlphaGo is based on three pillars: *deep learning*, *reinforcement learning* and *self-play*. In and of themselves, these are not new – what was innovative was the ingenious way in which they were combined.

Step 1: Deep learning

Deep learning has its origins in artificial neural networks (ANNs). These were inspired by the functioning of the human brain and represent an important trend within AI. The artificial neural network is a software model that, like our brain, is made up of neurons, each of which has several inputs. Different weights are given to each of these inputs to generate an output. By connecting neurons in a network, ANNs are able to learn complex relationships.

In the 1950s, when AI was in its infancy, the logic-based approach still predominated. Expert systems represent one type of this approach, with Mycin, a system for the diagnosis of infectious bacterial diseases, being a textbook example. In addition to making a diagnosis, Mycin was also able to recommend a treatment. An expert system consists of a set of 'If ... then ...' rules, which may be supplemented by uncertainties. Using a transparent reasoning process that is fairly easy for humans to follow, a conclusion is drawn about the possible cause based upon data such as symptoms. In addition to medical applications, expert systems were developed for the fields of aerospace, education and railways as early as the 1980s. Even back then, examples of expert systems could be found in the legal and military spheres.

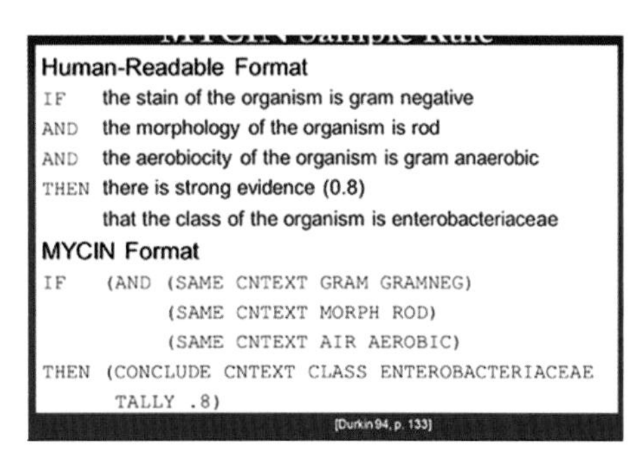

Doctor robot

Mycin, an expert system for medical diagnosis, at work. Mycin uses symptoms to determine a cause and suggest a treatment. The system does this in a way that mimics the human reasoning process.

At the time, the sub-symbolic approach used by ANNs was still in its infancy and had not yet achieved any real successes. Since then, sub-symbolic techniques have gained a huge amount of ground, with the use of ANNs becoming very popular in the 1980s.

Yann LeCun's successes in the field of automatic character recognition were a real milestone for gaining faith in such an approach. Suddenly, it was possible for a computer to recognize the numbers and letters that a human being had written down. This breakthrough was mainly due to the insight that ANNs, just like our brains, needed to be made up of multiple neurons combined in non-linear layers.

Another ground-breaking achievement was the BackPropagation algorithm that is still very popular today. That algorithm learns the parameters – or weights – of the ANN. During the training phase of the ANN, the weights are adjusted so that the input is optimally mapped to the corresponding output.

For example, take a camera image from a self-driving car. Every pixel of that image is fed into the network as an input. The output is then the desired steering action. To allow the network to learn, the network can be built into a car that is initially driven by a human being. The network is thus supplied with inputs (camera images) and the desired outputs (human steering actions).

To start with, the network will make mistakes and propose different actions to those taken by the human driver. It learns from these errors and adjusts the weights accordingly.

When such a network has been successfully trained, ANN can decide upon a sensible steering action even in the case of an as-yet unseen input, i.e. a new camera image. This is what we call the exploitation phase of ANN, in which the model has to demonstrate its predictive power.

LeCun also laid the foundation for deep ANNs. But the real breakthrough for deep learning was achieved only recently. A deep ANN is made up of multiple layers, usually seven or more, with each layer playing a different role. The lower layers look for a suitable representation of the input, whilst the upper layers solve the actual task.

In 2011, IBM's AI programme Watson beats its opponents in the American television quiz Jeopardy. Watson is a small data centre with some 2,280 processor cores and fifteen terabytes of memory, spread across ten server racks. It was not allowed on the internet during the game. It was represented by a screen.

©IBM Watson

From lines to a cat

The lower layers of a deep neural network search for basic elements, such as line sections. Together, these lines roughly make up the eyes of a human being or cat's ear. These elements are then combined in the upper layers. This makes it possible to automatically classify images, for example by determining whether or not the image contains a cat. It is also possible to search for images of an individual person.

Training these deep networks calls for not only a large amount of data (big data) but also a lot of computing power and energy. And these things are just what Google's DeepMind has access to in abundance. The first step in the development of AlphaGo was therefore the training of a deep ANN based on millions of moves by expert Go players. After this initial deep-learning stage, AlphaGo was able to imitate experts. But it was not yet ready for the big battle.

Example of a deep neural network

The lower layers of the network search for basic elements, such as line segments. Together, these line sections make up elements such as human eyes or cats' ears. These elements are then combined in the upper layers. This firstly permits the automatic classification of images – for example whether or not the image contains a cat – and secondly makes it possible to search for images of an individual person.

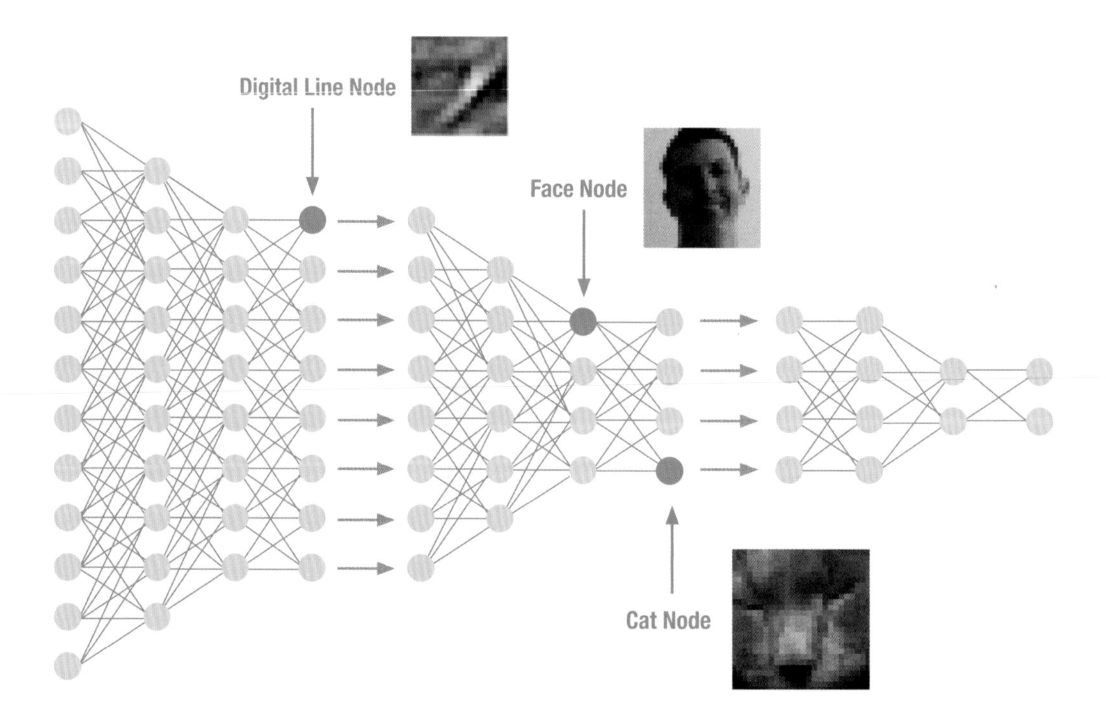

Step 2: Self-learning systems

In order to further improve AlphaGo, Google DeepMind drew upon reinforcement learning (RL), a method related to 'operant conditioning' in psychology. In this approach, desired actions by people or animals are rewarded – and undesired ones punished – in order to gradually teach the desired behaviour. This process can also be translated into an algorithm known as reinforcement learning, or simply RL, within AI.

How a rat learns to operate a lever

In an experiment, Edward Thorndike placed a rat in a small cage with a lever in it. When the rat touched the lever, food was released. Because rats see food as a positive reward, they are motivated to perform actions leading to that positive reward – in this case touching the lever.

RL is an appropriate method if the desired outcome of the task is known, but it is not clear in advance how it is to be achieved. Think of riding a bike. You can't tell a child exactly how to do it, but you can certainly indicate that the intention is to stay on the bike without falling over.

In a game like Go, the aim is to win a match. The 'best actions' are not specified; a reward is simply provided if the game is won. It is important to bear in mind that the victory is not necessarily the result of the last move: an earlier move might have been the brilliant breakthrough and it is this that must be rewarded. Similarly, a bad move may inevitably lead to the game being lost, but this may not become apparent until a few moves later.

In order to learn, it is also necessary to try out new actions with as-yet unknown outcomes. But, of course, the aim is to improve over time, not simply to carry out a long sequence of random actions. To achieve this, we need to find a balance between exploration – testing new actions – and exploitation – using previously acquired knowledge. In RL, this is called the exploration/exploitation dilemma.

RL in its purest form assumes no knowledge of the problem to be solved, only by indicating the actions that can potentially be carried out. However, this doesn't mean we can't speed up the learning process by incorporating knowledge from experts. This can be done in different ways. One is to enrich the description of the objective by providing hints about the learning process. For example: it's a good idea to maintain a certain degree of momentum when cycling.

Another is to encourage the learning process through comments such as 'you did that well' or 'we think that was a good move', or by demonstrating actions.

AlphaGo is based on the latter principle. It starts from an ANN that was initially trained with the help of expert players. An example from a different context is a robot learning to play table tennis. A human helps the robot arm to get started by demonstrating the movement a number of times.

But RL not only allows you to imitate the experts, it also helps you surpass them. To improve using RL, AlphaGo had to be given the opportunity to challenge an opponent. Initially, DeepMind invited the European Go champion. But, because AlphaGo had already reached quite a high level, it won those games fairly easily. That's where the concept of *self-play* came in.

DeepMind, which was acquired by Google in 2014, developed the AI system AlphaGo that hit the headlines all over the world in 2016 when it became the first computer to beat a professional player at the Japanese strategic game Go.

©Google

Step 3: Self-play

Self-play is a useful approach to strategic games such as Go. The RL has been given the goal of winning the game but, of course, it needs an opponent to play against. That opponent has to respond to the moves made by the AI player. Because it is not practical to use top-level human Go players throughout the entire learning process, and because the configurations that crop up during play are not necessarily included in the sample moves, the AI player plays against itself – or, rather, a copy of itself. When the learning AI player gets to the point where it can easily beat itself, it's time to make a new copy and continue the process. This allows the AI player to constantly improve using the RL learning process. This is how RL can ultimately surpass even the world champion.

This principle was neatly demonstrated in the second game against the world champion, when AlphaGo surprised Lee Sedol with an extremely creative move. The commentary was as follows: 'The Google machine made a move that no human ever would. And it was beautiful. As the world looked on, the move so perfectly demonstrated the enormously powerful and rather mysterious talents of modern artificial intelligence.' But precisely because AlphaGo challenges people through its creativity, Lee Sedol was also able to transcend himself. In the fourth game, it was his turn to surprise AlphaGo with a brilliant move, a move that AlphaGo had in no way expected from a human being.

So are robots now smarter than humans?

Beating the eighteen-time world Go champion must count for something. But does this mean that robots are becoming smarter than people? Go is both the simplest and the most abstract of all board games. It is simple in the sense that the rules are clear. There is no ambiguity about which moves are possible and when the game has been won. Unlike football, there is no need for a video assistant referee. But it is also abstract in the sense that it is very difficult to put a strategy into words. These are characteristics that lend themselves to AI and, in particular, to an RL approach. But there's more.

The fact that this is a board game in which the aim is to surround the opponent and thereby conquer territory means that deep learning is a highly suitable approach. Deep learning has made a particular name for itself in anything to do with image recognition, such as self-driving cars and medical image processing for tumour detection: a field in which AI can now outperform medical experts. Deep learning has also achieved many successes in natural language processing. But that's pretty much it. The building blocks of deep ANNs are optimized for such kinds of applications, and it remains to be seen whether we will be able to find the right basis for solving other tasks.

It is also important to realize that AI solutions are good at very specific tasks, but that transferring that knowledge to another context remains a major challenge. AlphaGo cannot simply use its expertise to play chess, distinguish between a dog and a cat, or enter into a simple dialogue with human beings. We have robots that can fold towels, but don't ask them to sort your laundry. And IBM's Watson, the champion at Jeopardy, can assist you with basic legal advice or answer simple medical questions, but it relies upon an AI approach that won't get it very far with Go.

Conversely, AlphaGo can't help with legal or medical problems. The AI techniques used are completely different, as are the problems to which they lend themselves. Although self-driving cars may provide a neat demonstration of the way in which the various AI components can be combined, their knowledge, once again, not easily transferable. And this is exactly something in which human beings excel. Humans have highly efficient learning processes that are not only useful for a specific task, but can be applied flexibly and with great skill to the many different challenges we encounter.

Man and AI as the perfect team?

Although AI clearly still has its limitations, the victory of AlphaGo has once again demonstrated its power. The European and world champions have both declared their deep respect for the AI player, with both admitting that the experience helped them reach a higher level in their own game.

So this seems like a case of 'man and machine as the perfect combination', as Kasparov put it when he was forced to accept defeat by IBM's chess computer Deep Blue. But there is still one piece missing from the puzzle. Although AlphaGo is able to learn a strategy, rather than relying upon raw computing power alone, it cannot explain the strategy it has used. AlphaGo cannot yet act as a real mentor to those learning to play Go.

An important step for achieving a better synergy between man and machine would therefore be to unlock the strategy being learned, making it transparent to human players. Such transparency could be the key to a better transfer of knowledge from one domain to another. This would mean that man and machine could learn in symbiosis, allowing a real relationship of trust to form. And the new GDPR privacy regulations do, in fact, impose a requirement of transparency.

WOULD YOU SPOT A PAINTING BY A ROBOT?

By Prof. Dr. Ann Dooms

The production of an artwork is driven by the senses, the brain, and the emotions of its creator. The digital age has given robots both senses and a brain. Sensors allow them to see, hear and feel, and a computer or computer network allows them to think, learn and make decisions. They also have 'muscles' – motors that allow them to execute movements. So, can a robot use these features to produce art?

How does a robot look at a painting?

A robot 'looks' at the world through cameras, which provide it with a digital image or video. That image actually consists of a very large table of numbers, which usually lie between 0 and 255, describing the amounts of red, green and blue light measured by the light sensors at that particular location in the camera. Lights of that particular colour and intensity light up on the screen of a TV, computer or smartphone, allowing us to view the image. We perceive this as a block of a certain colour. These blocks are known as pixels – a contraction of *picture* and *elements*. If you zoom in on a photo, you will see the pixels that make it up. There are usually 256 possibilities for red, green and blue per pixel. The rich colour palette of so-called RGB images is quite a good approximation of our reality. A camera actually imitates the workings of our eye, which also converts light into millions of electrical signals that are transmitted to our brains. When these signals arrive, they are converted into images, which are interpreted instantly.

So how can the robot's 'brain' allow it to 'see'? Its computer can analyse images by performing mathematical operations on the numbers, a procedure known as *image processing*. By detecting differences between successive pixel values, it can easily and quickly detect a sudden transition between small and large numbers, allowing it to perceive edges and shapes. Interpreting what it is viewing at any given time (the field of *computer vision*) is another matter. This is one more of the great digital challenges for artificial intelligence.

A robot looking at a painting can thus make out shapes ranging from a small, pronounced brushstroke to the composition of complete figures. It can use this ability to learn the style of a painter right down to the last detail by *machine learning*. To do this, we show it a series of paintings by the same artist, for example Vincent Van Gogh. It's best if these are displayed at the same distance and in the same lighting conditions, so that the computer can compare all the digital image information correctly.

There are also cameras that can measure more than just visible light in the electromagnetic spectrum – for example x-rays and infrared radiation. This allows the robot to detect information about things that is invisible to the naked eye, from sketches hidden beneath a layer of paint to air bubbles in the varnish layer. Its computer then performs various mathematical operations on all these images to study the edges in the painting, and the relationships between them, in greater depth. This allows the lengths and directions of patterns to be mapped out so that the computer can learn the artist's specific style, albeit in a series of numbers.

Sherlock Robot

A robot can only really be said to know the hand of an artist when it can use this series of learned numbers to detect fake paintings. The Van Gogh Museum in Amsterdam is still in doubt about the authenticity of a number of works in its collection, which is why this type of digital style analysis has attracted the attention of the conservator. She was sceptical about the technique and so, a few years ago, set up a challenge: a restorer was asked to make a copy of an existing work by Van Gogh (restorers are, of course, masters in painting 'in the style of'). She realized that the indiscriminate copying of a work of art would give rise to flaws in the brushstrokes, like the flaws when one tries to forge a signature. So, she copied the exact composition of the work, but coloured it 'in the style of' Van Gogh. But her work was detected by mathematical analysis due to her combination of brush stroke lengths and directions, which did not correspond precisely with the patterns learned from Van Gogh. It should be noted, however, that the style of Van Gogh lends itself perfectly to this method of analysis, largely due to his prominent brushstrokes.

Not every work of art can be studied digitally with equal ease. Just think of the Flemish Primitives, who used tiny brush strokes. Moreover, in that era a painting was the work of an entire studio: the master was not the only one to have a hand in it.

THE SEARCH FOR CRAQUELURES

A few years ago I was proud to visit the Royal Museums of Fine Arts in Brussels to present our 'Sherlock' technique. I thought they might have been interested in this. But nothing could have been further from the truth. Museums rarely have enough paintings by a particular artist to allow our algorithm to learn their painting style in detail if there is any doubt about the authenticity of a new purchase. However, by discussing the problems they face, we finally came up with the idea of using our mathematical techniques to determine – automatically and digitally – the conservation status of a work (craquelures, missing paint, underpaintings, etc.).

We have also been able to apply these techniques to the Mystic Lamb by Van Eyck. Of course, the computer is no substitute for an expert in these analyses, but it does give them an extra tool to aid their investigations or decisions. Using the digital painting style of the work or artist, the computer can also make predictions about what a missing piece of the painting might have looked like. And perhaps, in the future, a robot may be able to provide physical assistance to the restorer as they perform delicate operations.

Vincent Van Robot

If a computer can characterize the style of a painter in numbers, this naturally also opens the door to the production of digital forgeries. For example, *DeepArt*'s software already allows a photograph to be converted into a digital painting in the style of a single sample. Its *deep learning network* can separate content and style, allowing the style to be transferred from one photo to another without really affecting the content.

Using mathematics, a robot can thus mimic the style of a painting. If you were also to provide it with a relief image and the chemical composition of the original painting (information that you can even obtain from images using the x-ray fluorescence technique), this also gives the robot an idea of how to create its own digital painting in real life. It does not lack a steady hand, so there's nothing stopping you from having it actually paint its own creation.

DeepArt.io can convert a photo of the well-known robot Pepper into a self-portrait in the style of Vincent Van Gogh. It even automatically gives the robot a beard.

Industrial robotic arms have already been developed for creating art. In 1973, the painter Harold Cohen began work on his robot, Aaron. At the time, it was not yet possible to show the robot photos and have it learn patterns, so he manually coded it with the relationships between edges that give rise to natural forms (such as plants and people). His painting robot began to work with this data from the 1990s onwards.

Robotlab's Kuka robot looks at a model with its camera eye, mathematically analyses the shapes as described earlier, and uses this information to draw a portrait with a pen in its hand.

Machina Artis 3.0 by Carlos Corpa consists of several robots that together paint an area of 20 square metres to the rhythm of music (played by robots!). Leonel Moura's RAP (Robotic Action Painter) drives over a canvas while its coloured markers move up and down based upon mathematical calculations of the colour information it sees at that spot. This robot even signs its paintings.

Then there is the recent Scribit, a writing robot for interior decoration that is able to make an erasable drawing on walls and windows with ink that evaporates when heated. Why not decorate your own walls with robot art? What are you waiting for?

If you prefer something to hang on your wall, 3D printing is an option. In mid-2016, Amsterdam-based bank ING unveiled a 3D printed painting by Rembrandt, created by a computer that had analysed almost 350 digitized works (including relief information) by the artist.

It was then instructed by an algorithm to create a portrait of a man between thirty and forty years old, looking to the right, in dark clothing and with facial hair. Using facial recognition, the computer searched for paintings that matched that profile to work out what a typical eye, ear, nose, mouth, etc. by Rembrandt looks like. It also studied shadows and textures, so that it could use all of this information for its composition (if we can call its work that – perhaps it is

A Scribit robot at work.
©Scribit

better described as simply the average of what it has learned). The result, made up of 148 million pixels, was created by a 3D printer using special paint-based ink that imitated Rembrandt's brush strokes in thirteen layers.

Do robots produce 'real' art?

In 1950, the English mathematician Alan Turing published the article *Computing Machinery and Intelligence.* This article opens with a thought experiment he called *the Imitation Game.* In one room is a human 'interrogator' and in the other a person and a computer. The interrogator is given the task of finding out which of the two is the human being through a series of typed questions that are sent back and forth. The question of whether the computer can fool the interrogator is known as the *Turing Test.*

That simple but powerful thought experiment actually provides a general framework for studying the interaction between man and machine. The Dartmouth College Neukom Institute for Computational Science has put this into practice by developing a *Turing Test in the Creative Arts* into which anyone can enter their robot. The robot-generated works are put alongside

The Next Rembrandt, designed using AI and produced by a 3D printer.
©J. Walter Thompson Amsterdam

material created by humans and an (extensive) jury assesses whether each work was created by a robot or a human being. The winner is the robot whose work is statistically indistinguishable from the human material. Cohen claimed that his Aaron robot would pass the Turing test. However, we should note that the robot – like the one that produced *The Next Rembrandt* – was only creative within the limits imposed upon it.

The Lamb robots

If we let DeepArt loose on a photo of the VUB campus and the Mystic Lamb by the Van Eyck brothers, the outcome is rather bizarre. Could a human being have created that work? The Van Eyck brothers' photorealistic painting style makes it difficult for the deep learning software to separate content from style, leading the computer to use figures as texture for the tree. That gives rise to a result that is interesting, to say the least.

The Turing test can be applied to more than just robotic painters. Iamus, an algorithm that composes classical music, was created at the University of Malaga and has had two of its compositions performed by the London Symphony Orchestra. The first violinist was impressed, but felt that the music didn't 'go anywhere'.

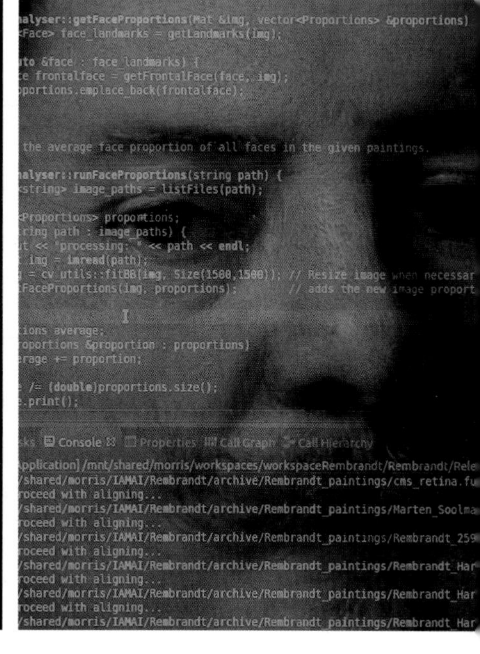

A 'Van Eyckification' of the VUB campus

A similar experiment was set up by VRT (the national public service broadcaster for the Flemish Region and Community of Belgium) and Robovision. A computer learned the style of pianist Jef Neve from a few hours of improvisation. The computer then created a new composition in the same style. Neve himself said he could hear 'a sequence of very interesting words, but no sentences'. These two musicians hit the nail on the head about what seems wrong with the 'Van Eyckification' of the VUB campus. It thus remains to be seen whether a robot will be able to fool an art connoisseur at some point in the near, or more distant, future.

In his book *Homo Deus*, Yuval Noah Harari claims that pattern recognition is all it takes to perform a large number of our jobs. But things seem to be more difficult in the case of art. Does creativity, too, call for the ability to interpret patterns? Does a robot need to have emotions to make 'real art', or is it enough that its work evokes emotion?

In the meantime, the work of robots will, in any event, give rise to new forms of art just as, a hundred years ago, the painter Wassily Kandinsky was inspired to create abstract art by developments in mathematics. Perhaps robots will become companions to artists and they will work together in the studio on co-productions between man and machine? Perhaps creating an autonomous robotic artist is the ultimate art form. A work of art that speaks the language of mathematics.

WILL A ROBOT BE YOUR BEST FRIEND?

By Dr. Werner Schirmer

There is yet another home robot. It looks like a cross between a small excavator and a forklift truck, and has two very expressive, square, green eyes. Its name is Vector. In a promotional video, the manufacturers boldly advertise their product as a 'big milestone for robotics because for the first time ever there will be hundreds of thousands of people living with a robot in their homes'. They promise that you will 'feel something for him, bond with him, relate to him, care about him' and that you will feel 'that he can relate to you and give you something back'. They also think he represents 'a step closer to the sci-fi robot that everyone wants to be friends with'. One of Vector's main functions is to 'hang out' with you.

That description of Vector is not as far-fetched as you might think, but we are still a long way from the utopian social robots we know so well from science fiction films and TV series. Bender from *Futurama*, Number 5 from *Short Circuit* and C-3PO from *Star Wars* are well-known examples of robot friends who remind us of anthropomorphic creatures but still look like mechanical machines. Ava from *Ex Machina*, the 'inhabitants' of *Westworld* and the 'real humans' from the series of the same name look very human, and behave in a very human way, which could make it easy to forget that there is a lot of technology hidden behind their appearances. Nonetheless, these fictional robots have contributed to the hype and growing expectations around social robots.

What are social robots?

First, let's take a brief look at what we actually mean by 'social robots'. As the other chapters in this book show, not all robots are social. Many industrial robots and autonomous household appliances (such as vacuum cleaners and lawnmowers) are designed to perform repetitive tasks for people. Social robots, on the other hand, are designed to interact with people. One type of social robot is designed to meet instrumental objectives, such as delivering mail in an office or serving customers in shopping centres or hotels. An example of the latter is the robot Mario who has been working at the Marriott Hotel in Ghent since 2015.

The other type of social robot is designed to fulfil affective goals. One of the first such social robots is 'Kismet', developed in the late 1990s by robotics engineer Cynthia Breazeal. She defined affective, social robots as 'able to communicate and interact with us in a personal manner, and to understand us'. Among the social robots developed by researchers from the VUB, the best known is called Probo.

Care and companionship

There are two main paradigms to affective social robots: the *caretaker paradigm* and the *companion paradigm*. In the *caretaker paradigm*, it is human beings that take care of robots. They have to keep the robot 'satisfied' by identifying its emotional and social needs and responding appropriately. The Tamagotchi of the 1990s is one of the best-known examples of this phenomenon, although it was more like a virtual pet in the form of a plastic toy egg than a robot. Other, more recent, examples are the baby seal Paro and the robotic dog Aibo, which can be stroked and are able to express emotions such as anger, sadness, fear and happiness.

The main activity of robot buddy Vector is to wander around on a table and explore its surroundings. The social robot is also able to play games, set timers and answer questions.
©Anki

In the *companion paradigm*, on the other hand, it is the robot that takes on the role of caregiver. Its job is to identify the needs of human beings – for example lonely elderly people in residential care homes – and respond appropriately. Robots for use in such situations must be designed so that human users feel comfortable with them. So, the robot needs social skills, plus the ability to adapt to the user's personality, likes and dislikes.

Why would we want a robot as a friend?

Now, why would we want to bring a robot into our homes to act as our companion in the first place? The social sciences tell us there are several reasons. Human beings have an innate need for connection, and a strong desire for meaningful interpersonal relationships. Thanks to an evolutionary mechanism that helped ensure the survival of our ancestors over 50,000 years ago, we feel sad and down if we become too disconnected from our group. Prolonged loneliness can lead to depression and serious physical illness.

Sociologists have pointed out that modern society is fundamentally different from traditional societies. Social cohesion and the sense of community used to be much stronger, but so was social control and the suppression of deviant behaviour and opinions. Today, we tend to treat our fellow human beings in a rather blasé fashion. The 'individual' is a modern invention that has opened up great freedoms, allowing us to become emancipated and to escape social control and oppression. However, there is a price to pay for this individualism. Disconnected from our roots in our community and family, we have to fall back on ourselves. As individuals, we need to function and perform in all kinds of competitive markets. The strong sense of connectedness and identity throughout the entire human lifespan that characterized past societies has given way to patchwork identities and a constant struggle to avoid loneliness.

Is modern technology turning us into narcissists?

In recent years, the rise of social media has reinforced the trend of individualization. Although they preach connectedness and community, social media platforms ultimately breed a cult of the individual, of self-promotion, social benchmarking and personal branding: the individual as a product. This turns people into narcissists lacking in self-confidence and constantly yearning for social proof and confirmation through *likes* and a growing number of *followers*.

Psychologist and robotics expert Sherry Turkle has noted that young people today are afraid of face-to-face interaction and are therefore opting for less 'intrusive' or confrontational methods of communication. They are less and less able to express and interpret feelings without the help of emoticons.

Against this backdrop, it seems plausible that we are increasingly expecting more from technology and less from each other, as Turkle puts it. Mediatization and the ubiquity of smart technology in everyday life are making personal interaction with 'real' people less and less important. Interaction with bots (in virtual worlds and on social media) and robots , by contrast, appears to be less problematic. We could even argue that the prospect of artificial friends seems more attractive than that of human ones because robots, as designed machines, will always be there for us when we need them. They listen to our whining and complaining all day long (or at least until their battery is flat) and don't bother us with their own emotions or mood changes (on the contrary: in the *caretaking paradigm* we consider them 'cute'). This means we can project our fear of loneliness and our hopes regarding the fulfilment of our social needs onto robots.

Not surprisingly, Turkle also reports on a few young women who would rather have a robot than a real man as a boyfriend, because social robots promise companionship, but don't demand much in terms of emotional maintenance.

What does it mean to be social?

In order to define what makes social robots 'social', we first need to understand what the word 'social' actually means. A central insight of sociology is that the word 'social' refers not to attributes of attributes of people themselves, but to something *between* people as they interact. That 'something' emerges through symbolic interaction and manifests itself as an arrangement of generalized expectations. By 'social structures' we mean different levels of generalized expectations that people use to orient their actions: rules for interaction (for example, how to behave in a job interview), complementary social roles (parent-child, teacher-pupil, salesperson-customer, or footballer-referee) and culturally shared beliefs and norms.

In principle, these expectation structures are changeable – our society, for example, has become much more pluralistic and liberal than it was a hundred years ago. Expectation structures are also challengeable, although such challenges are often followed by embarrassed or angry reactions. Yet most social scientists agree that the social environment in which we grow up largely defines who we are, what we like, how we express ourselves, and so on. We are born into existing social structures and are 'programmed' by social institutions such as family, religion and schooling. While society as a whole is constantly changing, most of our social lives follow fairly predictable routines.

What makes robots social?

Now we could argue: If people are programmed and follow routines, and the same applies to robots, what makes us so different? And why is it so difficult to make robots social? Why can't we just program robots with behavioural norms, rules, social roles and the like, and design them to behave in as socially acceptable a way as possible?

There are two problems with this idea. The first is related to culture and power. Not only is there a lack of agreement about which core values are most acceptable within a certain culture (for example freedom of speech versus social justice), but different value systems also exist across cultural boundaries (individualism versus collectivism, women's rights versus religious authority). While people are 'programmed' locally through a range of social institutions, robots are designed, and thus programmed, centrally by manufacturers. If manufacturers are not controlled by the government or legislation, they (and their shareholders) could gain enormous

The social robot Elvis uses its arms to make various gestures and exhibit a range of emotions.
©VUB-Brubotics

power to define or redefine what constitutes socially acceptable behaviour, similar to the power enjoyed by internet platforms such as Facebook, Amazon and Apple.

If social robots were to move into our homes *en masse* – as predicted in the promotional video for Vector – robot designers would increasingly be confronted with conflicts between commercial, political and ethical interests. Would robots programmed with a single set of contemporary values give rise to conservatism and prevent social change?

The second problem is that, even though robots outperform human beings at some tasks, they are also clumsy when doing certain jobs that people find easy. Social competence is one of the things that comes naturally to most (though not all) of us, but is very difficult for robotic engineers to model. Although speech recognition and synthetic speech are improving all the time, and conversations with AI assistants like Siri and Alexa feel almost realistic, those devices don't understand language in the way we humans do. That makes them notoriously bad at interpreting different layers of meaning, such as irony, allusions or humour. They also face difficulty in distinguishing between what is relevant and what is not, and struggle to adapt to changed social contexts. Because of these limitations, users often find interacting with today's social robots disappointing, boring or even irritating.

Emotions: real or simulated?

Advocates of the 'strong approach' to social robotics argue that true social robots must be able to express and recognize emotions, natural gestures, personality and identity. This means they need representative models of the user and themselves. As of today, even such a task is very difficult. Because of these technological challenges, supporters of the 'weak approach' prefer to promote 'socially evocative' robots, which are designed to prompt social behaviour on the part of the user.

Research shows that many users don't care whether a robot is truly emotional and intelligent, as long as they get that impression when interacting with it. This means that a very primitive level of emotionality and intelligence is sufficient. The classic example is Eliza, a computer program that mimics a psychiatrist developed by Joseph Weizenbaum in the 1960s. Although users were fully aware that they were interacting with a 'dumb' computer, they felt they were talking to a real therapist. Children and older people who interact with the robotic animals Paro and Aibo also experienced companionship and perceived the machines as 'almost alive'. When correctly designed, robots can actively arouse feelings in us and make us want to take care of them, whether they look like seals, forklift trucks, or flowerpots.

Interestingly enough, this connection is less clear for humanoid robots. On the one hand, robots with more human characteristics are perceived as more attractive. On the other, if the imitation is highly accurate but just not perfect, we perceive them as unnatural and creepy – robotics experts call this 'the uncanny valley' effect.

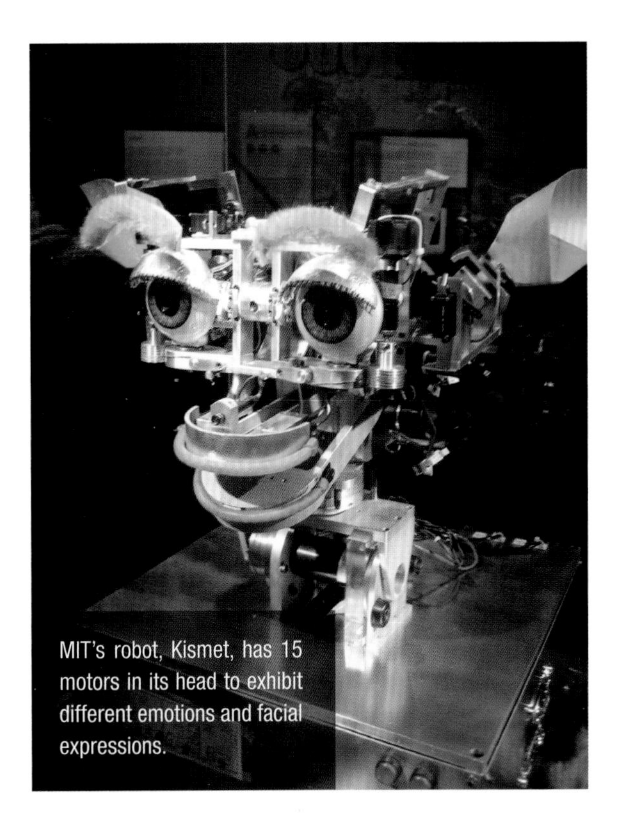

MIT's robot, Kismet, has 15 motors in its head to exhibit different emotions and facial expressions.

Ultimately, it comes down to whether we are satisfied with synthetic emotions and imitated sociality. Do we really want to form an emotional bond with machines that can't feel and have no idea what feelings are? Critics such as Turkle fear that vulnerable groups (children, lonely older people, people with disabilities) will be deceived. But others consider this approach to social robots as the right path, at least until the technology becomes more sophisticated. Their argument: As long as people enjoy their robot buddy and don't get bored or angry with it, why not?

Although we are still a long way from the robotic pals depicted in science fiction films, our world is gradually becoming saturated with gadgets such as Vector, and increasingly populated by social robots. The sociologist Zhao is convinced that, over time, the presence of social robots will transform the nature of social interaction, society and the individual. It is the joint task of robotic engineers and social scientists to study how, and to what extent, this happens.

THE TWO FACES OF A SOCIAL ROBOT

The Caretaker

Principle:	The human takes care of the robot.
Purpose:	To keep the robot 'satisfied' by identifying its emotional and social needs and responding appropriately.
Use case:	Toys, pet substitutes.

Characteristics:	The robot can express emotions such as anger, sadness, fear and happiness. The baby seal Paro and the robotic dog Aibo can be petted and react accordingly.

The Companion

Principle:	The robot takes care of the human.
Purpose:	The robot identifies the needs of the human being and responds appropriately.
Use case:	Companionship for lonely older people in residential care centres or for children with developmental disorders.
Characteristics:	Needs to be designed appropriately so that the human user feels comfortable with it. Must have social skills and the ability to adapt to the user's personality, likes and dislikes.

The sci-fi robot that everyone wants to be friends with and that 'hangs out' with you needs to have characteristics of both, but as of yet the technology is still a long way from this point.

MIT's robot, Leonardo, is one of the most expressive robots in the world. Although the social robot is not human, developers have deliberately given Leonardo a youthful appearance to encourage people to interact with it playfully, as they would with a young child.

©MIT - Personal Robots Group, MIT Media Lab

WILL YOU FALL HEAD OVER HEELS IN LOVE WITH A ROBOT?

By Charlotte Jewell and Prof. Dr. An Jacobs

A travelling companion for adventures in faraway places, a soulmate for long walks on the beach – or just someone to chill out with in front of Netflix: for many adults an intimate relationship is an important part of life. So, imagine having an intimate relationship with a robot. This probably sounds highly futuristic, bizarre and disturbing. Wouldn't this blur the norms for 'good' intimate partner relationships, leading to the impoverishment and erosion of social relationships, and to an increase in sexual abuse and violence? In 2008, David Levy published his book *Love and Sex with Robots*, in which he tries to explain the origins of our fascination with sex robots. He thinks that we will soon take marriages between a human and a robot in our stride. In his opinion, the social changes that are taking place in our society with regard to relationships (for example same-sex marriage, mixed-race relationships, etc.) and the ways in which we think about relationships will soon take us to that point.

Our imaginations running wild

Think how many robots you have encountered in real life to date. You may have seen a demonstration in a museum or other public place but, for most people, robots are not a part everyday life. In fiction, however – for example in science fiction films and TV series – the opposite is very often true. Such works portray a whole range of friendships and relationships with robots.

How the development of relationships with robots is portrayed depends largely on the period in which the film was made, with films essentially providing us with a snapshot of history. In the 1970s, the search was on for a utopian, harmonious human-machine relationship, as depicted in *Star Wars*. In the 1980s and 1990s, the emphasis was on using emotions to highlight the differences between people and machines, due to the growing fear that machines would detract from human uniqueness – *Terminator* is a good example in this regard.

The interpretation, in our imaginations, of the future of intimate human-robot relationships in sci-fi is, in turn, a reflection of our expectations about the evolution of human-human relations at the time the film was conceived and made. In other words, we replicate the knowledge that prevails at that time and in that place.

The film *Bicentennial Man* (1999) portrays the development of this kind of intimate relationship: first the getting-to-know-you phase and then the establishment and development of a bond. The relationship ends with the death of the main character Andrew (a robot) and his wife Portia. Over the course of the film, we watch as the classical stages of adult heterosexual romantic relationships emerge. What we see in sci-fi films and TV series has an impact on the way we view

Sex robot Harmony in the making.
©Realbotix, realbotix.com

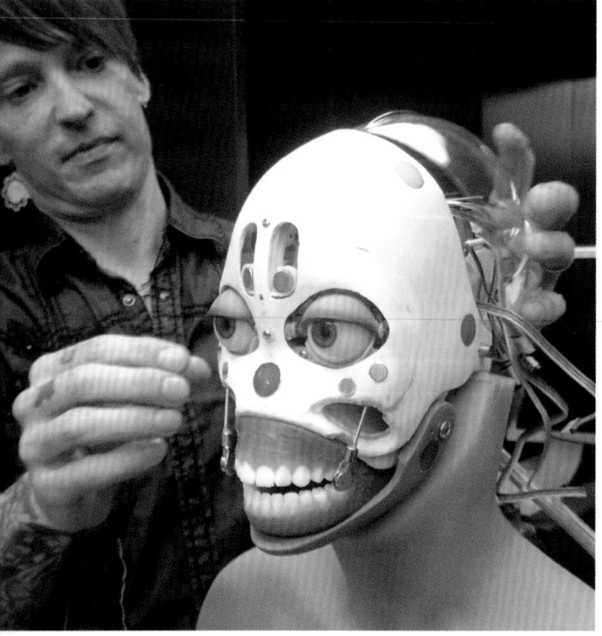

intimate relationships with robots, and what we expect from them. The sensational stories told by sci-fi films – which can have either a positive or a negative slant – influence our ideas about the future. As a result, we may fail to consider the more moderate scenarios. If we look at the tangible experiences that are currently possible, we may feel we are missing out on a range of opportunities.

Cradling a robot

It has been shown many times that we love to attribute human or animal properties to non-human objects (a phenomenon known as anthropomorphism) and to treat such objects as we would a human being. In the Wonder project, residents and staff of a residential care centre perceived a robot as a child, and treated it accordingly. Although the robot was remotely controlled, and therefore had only limited autonomy, people still formed a bond with it. Some residents rocked the robot in their arms as they would a baby, and spoke to it as if it were a child. Staff also avoided touching the robot in the places that would, in a human, be intimate sex zones, because it felt strange and wrong.

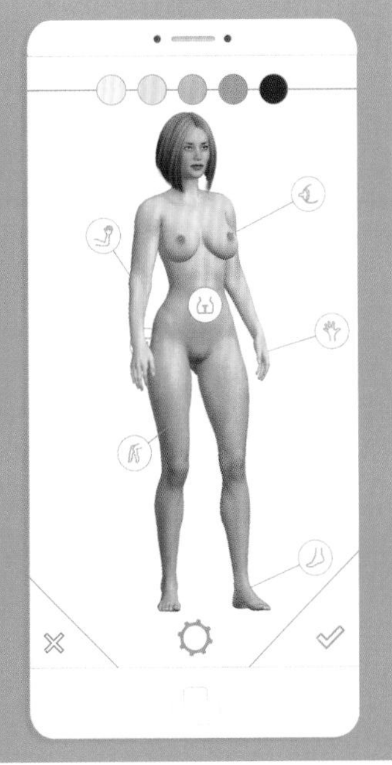

Busty robots

However, much of this technology is still in its infancy.

Robots currently being developed with the aim of creating someone's fantasy partner can be customized in appearance right down to the smallest physical detail in line with the tastes of the customer/future partner. Meanwhile, the search is on for ways to develop a tailor-made 'personality' – for example Frigid Fara, Wild Wendy, S&M Susan. But this calls for further technological developments within the field of artificial intelligence.

Central to this is the pornographic fantasy of the heterosexual male. The sex dolls produced are mainly women with bodies modelled on categories from the pornographic industry.

They are the more technologically advanced successors of the inflatable doll, which itself only became possible thanks to innovations in the field of plastics in the first half of the twentieth century, and was commercialized in the second half of the twentieth century. The archetypal doll that has been in use over the centuries has thus been given an update, but still conforms to patriarchal stereotypes. There are thus genuine concerns about the way in which such dolls belittle women and undermine their position. A recent American survey indicated that two-thirds of participating men saw themselves having sex with a robot at some time in the future, whilst two-thirds women did not envisage themselves doing the same thing.

The first steps towards the diversification of sex robots are being taken in the production of male specimens – such as the Real Doll – but, regardless of gender, the general type remains the same. They still aim to replicate a human being in both appearance (form and movement) and internal aspects (personality and style of interaction).

However, it would also be worth investigating other robotic manifestations so that we can come up with complementary functions for such a device – the comforter, the satisfier – rather than presenting it as a substitute. From the scarce research done on perceptions of this type of robot, we know that people see it as a problem for their partner relationship.

Thinking outside the box

Current developments in the domain of the robot as a fantasy sexual partner also raise other questions: Are we not in danger of blurring norms about what constitutes a meaningful relationship in our society (respect, mutuality) when people enter into relationships with robots? All too often, people may assume that they are entering into a mutual relationship, when this is purely an illusion. This can lead people to feel cheated when the illusion of reciprocity does not withstand the test of time. That's why it is important to keep playing with social signals and forms, so that there is no chance of mistaking the entity for a human or animal. Over time, new expectations may be linked to this – the robot should not respond in the same way as what we are used to.

There are already examples of developments that think outside the box in this way, such as Somnox, a bean-shaped robotic sleep cushion, or Sanwe's robotic sperm collector. When the form is less familiar from everyday life, expectations in terms of behaviour are also more limited. That's why Paro was deliberately made to look like a seal instead of, say, a cat.

Robotic sleep cushion Somnox
©Somnox, meetsomnox.com

Changing ideals as an inspiration

If we replicate what we are already familiar with in robots, how will our ideal for intimate relationships influence the development of future robotic partners? Let's take a closer look at that changing ideal.

When we talk of intimate relationships, we immediately think of sex. But there are many forms of love and intimacy, and you can be intimate without having sex. Having trust and respect for one another, daring to open up, being attentive to each other and cuddling are also types of intimacy. But let's stick to the forms of intimate partner relationships with a sexual component, which we will refer to as 'partner relationships'.

For social beings like humans, partner relationships are an important aspect of life. Most people feel that partner relationships give them a sense of meaning. Studies also show that healthy relationships have a positive influence on our physical and psychological wellbeing.

Yet we often forget that partner relationships have both a public and a private dimension. Through relationships, we are socialized from birth – it's how we learn the norms and values of the social group(s) to which we belong. Over time, these become completely self-evident, an integral part of who we are. Our relationships with our family, friends and partners remain primary reference points throughout our lives. The standards and values we have learned and internalized are thus constantly reaffirmed. These also influence the very personal choice of an intimate partner relationship. Although marriage has been redefined considerably in recent decades, the married heterosexual couple remains the dominant point of reference for meaningful intimate partner relationships in the West.

But studies also suggest that modern marriage can also be seen as a 'suffocating model'. Whereas, in the past, security and the fulfilment of physical needs were our main priorities, nowadays individual needs come first. We expect our partner to support us with personal growth, emotional intimacy and companionship and provide deep physical and emotional sexual satisfaction. This gives rise to a number of high expectations that cannot always be met because we are also spending less time on our marriages. This can have a detrimental effect on the personal wellbeing of both partners. We therefore need to adjust our expectations to bring them into line with what is actually possible, thereby giving the marriage more room for manoeuvre. The roles that each partner expects the other to fulfil – for example those of breadwinner, lover or activity partner – need to be open for discussion and prioritization. Defining the roles or needs that are not a priority within a marriage opens up the possibility of other people, or perhaps a robot, fulfilling these roles. Other, more controversial, relationship models also come to mind, such as long-distance relationships or *consensual non-monogamy*, in which a robot partner could play a role.

The search for a meaningful relationship

But how do we know whether we are developing robots that permit quality partner relationships? Note that we are talking about the 'quality' of a relationship and not about whether a relationship is 'good' or not. After all, what do we mean by a 'good' relationship? This is different for everyone. For one person, it might mean a monogamous relationship, while for another a polyamorous relationship is the ideal. We therefore prefer to talk about a 'meaningful' rather than a 'good' partner relationship. What constitutes meaningful is down to the interpretation of a person living in a certain time and place.

But, coming back to the main point: how do we assess quality? We can adopt the methods used in research and relationship therapy for human-human relationships to examine the quality of human-robot relationships. These methods can help identify couples in need of relationship support, or in understanding how relationships work. Relationships are studied based upon the degree of commitment, communication, conflict resolution, interaction, time spent together, intimacy and emotional support.

But there is a catch, because most of the methods used are designed for heterosexual, married couples. This does not necessarily make them suitable for the various forms of relationships we see in today's society, such as LGBT couples. There is still work to be done in that area!

Still plenty of scope for imagination

When developing social robots, care is therefore taken to create an appearance to which limited social expectations are attached, so that it's not necessary to meet high expectations. Aiming for the replication of human beings gives rise to a number of problems in the field of partner relationships. What are the options if we don't do this? Can it broaden our range of possible robotic partner relationships, and thus enrich our view of human-human partner relationships? There is still plenty of scope for imagination when we think of all the variations in partner relationships within our contemporary society.

META

Experiments have shown that people are less likely to switch the robot off if it begs them not to.

©Aike Horstmann, University Duisburg-Essenn

HOW ILLOGICAL DO ROBOTS HAVE TO BE FOR US TO LIVE WITH THEM?

By Prof. Dr. Jean Paul Van Bendegem

It's a well-known joke that, at one time, Windows users had to switch their computers off by pressing the 'start' button. This seemed a bit strange and illogical, to put it mildly. But was it? You could reason that the obvious place to finish is back where you started, i.e. at the 'start' button. But in that case isn't it a bit odd to call it the 'start' button? Why not 'on/off'? Surely that would be more logical? Maybe, maybe not. 'On/off' doesn't really tell you what the button does: the two commands are mutually exclusive. Actually, 'start' is the obvious choice, because it's the first action you perform when you boot up your computer. And if you know that ending means returning to the start point, you'll look for the 'start' button and not the 'off' button. Surely that sounds perfectly logical? Doesn't it? This article takes a brief look at the unease that many readers may have felt when reading the explanation above. It all seems to be true, but at the same time it seems strange and illogical. Why?

What we have done above is reasoning. We have taken a situation, obtained information and used this information to try and derive new information. We also want to assess the reliability of this new information.

Suppose a robot says that robots will soon take over the world. A human being asks how the robot knows this, and the robot answers: 'Because I said so'. Most of us will be inclined to switch that robot off. But if the same robot gives us a long-winded explanation containing economic, sociological and political arguments, backed up by data and diagrams, we may well be convinced – and we'll probably panic.

But – and this question gets right to the heart of the matter – why should we accept the information in one case and not in the other? In looking for an answer to this question we enter the field of logic or, as I like to call it, the art of quality reasoning.

There is no doubt that logic has an important role to play in robotics. A robot with sufficient autonomy should be able to derive new information from the information it has already been given. So, what's the best kind of logic to give the robot? Isn't it critical we ensure this information is as reliable as possible? Or would that make us too uncomfortable?

So what is logical reasoning?

This problem brings us to the question of what logic is. The oldest known example in our Western history, dating back to Aristotle, is a conception of logic in which the new information must be absolutely reliable. Example:

(P1)	If people are thirsty, they drink.
(P2)	If people drink, they have to pee.
(C)	If people are thirsty, they have to pee.

If we accept statements (P1) and (P2) we also have to accept the conclusion (C). But why? The generally accepted answer in logic – and I use this wording to highlight the fact that this is not the only possible answer – is that the reasoning is correct because it follows the following schema:

(P1)	If A then B.
(P2)	If B then C.
(C)	If A then C.

If we take A to be 'people are thirsty', B to be 'people drink' and C to be 'people have to pee', then you can see that the reasoning follows the schema. But why is that schema acceptable? Now we are getting to the heart of the matter.

It's very tempting to decide that a schema is acceptable by reasoning logically about it, but that would be a mistake. That logical reasoning is itself based upon schemas, and so we are justifying schemas with more schemas – and going round in circles. So, what arguments can we use?

There are lots of possibilities:

1 Because it is intuitively clear or obvious:

(P1) A and B.
(C) A.

If we accept that the sun is shining and that it is Thursday, then we must accept that the sun is shining.

2 Because there is an existing reasoning practice that corresponds to this schema:

(P1) A.
(P2) If A then B.
(C) B.

If you accept a statement A, you also have to accept the consequences of that statement. Suppose I challenge a boxer to a fight and I get beaten up, I would expect bystanders to say: 'You should have known that would happen!' They unconsciously applied the above schema.

3 Because it is a consequence of the meanings of the words we use. For example, if we can agree that 'not-A' means that A is not the case, then, if A is the case, it is impossible that not-A is the case. Or, expressed in a schema:

(P1) A.
(C) not-not-A.

The main outcome of all the above explanation is that we don't always agree on the acceptability of a particular schema. As a result, there are different types of logic and which we use often comes down to a matter of (justifiable) choice. A classic example of a disputed schema is this:

(P1) If A then B.
(P2) If A then not-B.
(C) not-A.

This schema is known as *reductio ad absurdum*. If we were to accept A, then we would accept both B – based on (P1) and schema (2) above – and not-B – based on (P2) and the same schema. That's not possible, so we reject A. And what's more: we accept the refutation of A. Is that always wise? Are we really going to make the following decision?

(P1) It's a good thing if I die (because then my family can finally inherit from me).
(P2) It's not a good thing if I die (because I will cease to exist).
(C) I won't die?

Logical reasoning in the real world

I wrote above that logic has traditionally been concerned with absolute certainty. Unfortunately, this is a situation we rarely encounter in the real world. Even mathematics doesn't always achieve it. We 'ordinary' people, in everyday life, often have to make decisions on the basis of incomplete, unreliable information. The good news is that logic in the modern era has tried – and still is trying – to express, and assess the quality of, even that type of reasoning. But again the variation is great. A few examples:

1. Non-monotonic logic: this type of logic takes into account the possibility that information (and by extension knowledge) may change. If I tell you that Jan lives in Flanders, is Belgian and religious, then it's safe to assume – though by no means certain – that Jan will be a Catholic. But if we also know that Jan makes regular visits to the mosque or synagogue, we will need to review that conclusion. But, even then, our conclusion is not absolutely certain because Jan may be going for the sake of friends or family. We come across this in many debates and discussions: we pass judgment on a situation, but additional information forces us to revise that conclusion.
2. Default logic: this type of logic falls within the previous category and closely resembles one aspect of scientific method, namely induction. One of the many forms of induction is arriving at a generalisation based on a finite number of observations. You see a raven and it can fly, you see a sparrow and it can fly, you see a seagull and it can fly – isn't it an obvious conclusion that all birds can fly? Perhaps, but often not. It is true, however, that, based upon everything you know, it is reasonable to make this generalisation. In other words, it is compatible with everything you know to assume that all birds can fly. Until you see a penguin and then we are back in category (1).
3. Fuzzy logic (or, by extension, multi-valued logic): this is undoubtedly one of the most popular approaches. Take people with beards, for example. There is certainly a group of people for which it is indisputable that every member of the group – Ian, for example – has a beard. The statement 'Ian has a beard' is thus assigned the value 1 (which stands for definitely true). There will also definitely be a group – to which Pete, for example, belongs – for which the opposite applies and so 'Pete has a beard' is assigned the value 0 (definitely false). But what about all the doubtful cases in between, for example Charles with his stubble or Jeff with the little goatee? In fuzzy logic, these cases are assigned values between 0 and 1. If the statement 'Louis has a beard' has a value of 0.99, then it is virtually certain that Louis actually has a beard. But if you say the statement 'Marcel has a beard' has a value of 0.5, then what you are actually saying is that you don't know. True and false lie at the extremes of a spectrum of possible truth values, which allows for many more nuances.

One thing is clear: we find ourselves faced, once again, with a wealth of possibilities. Here, too, discussion hinges on which system of thought is best suited for detecting new, reliable information. However, there is a major complication that we cannot and should not avoid.
It's one thing to look for quality schemas, but quite another for we human beings to use them effectively. In other words: do we think logically? Surprisingly enough, the answer turns out to be no. There is extensive literature on this subject, but Daniel Kahneman's recent book brings everything together nicely. A concise and incomplete summary of how illogical our thinking sometimes is: (a) we use reasoning schemas that are incorrect, (b) we have difficulty with long and complex reasoning, (c) we are often guided by and cling to our initial impressions, (d) we are biased, (e) we see connections where there are none, (f) we prefer to look for confirmations rather than refutations, (g) we are bad at estimating probabilities, (h) we think in terms of stereotypes rather than populations, etc.
The good news is that we are aware of all that – but that doesn't make all those curious characteristics go away. The thinking of homo sapiens is, in fact, a bit bizarre and you would do well to bear that in mind.

What kind of logic do robots need?

The argument so far has in fact been a long-winded prelude to the key question: what kind of logic do we need to equip a robot – or a homo roboticus – with? It seems that we are facing a dilemma – a classic schema that says that if there are two alternatives and both have the same effect that effect is unavoidable. This is because:

1. either we let the robot think in a logically acceptable way by using only correct thinking schemas;
2. or we have the robot think like a human being by adding a series of faulty schemas (and we should add that there are already studies that support this approach).

In the first case, the homo sapiens will not recognise the homo roboticus because its reasoning is so 'strange' and it comes up with conclusions that seem a bit off to us (like pressing 'start' to stop, for example). This kind of homo roboticus will evoke strange reactions at the wheel of a car. But in the second case, we are turning the homo roboticus into a pseudo-homo sapiens – and do we really want to let even more bad drivers out into the traffic? Put this way, it seems inevitable that we will have to expand the way we deal with 'different thinkers' from homo sapiens to homo roboticus. In other words, a genuine exercise in open-mindedness!

EYE CONTACT

At first sight, it looks ridiculous, doesn't it? Is it supposed to be a children's toy or is there something else going on here? There is indeed more to it than that: Jaguar Land Rover has developed this self-driving car with eyes because it has been shown that the vast majority of people are unwilling to cross the road in front of a self-driving car. Cognitive psychologists have found that eye contact with the driver is very important when we are trying to work out what he or she is likely to do next. But a self-driving car doesn't have a driver. So this car has been fitted with its own kind of eyes. If it stares at you, you know that the system has detected your presence and you are safe to proceed. Isn't this a wonderful example of mutual adaptation between man and machine?

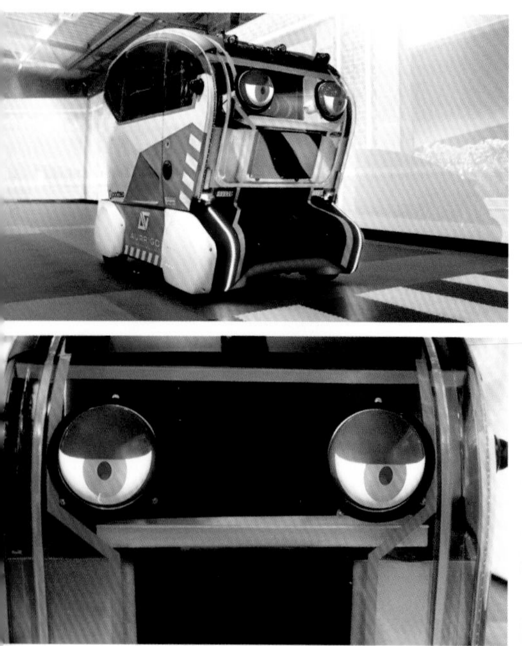

©Jaguar Land Rover

DO ROBOTS HAVE A BODY AND A MIND TOO?

By Dr. Johan Loeckx

In the summer of 2000, philosophy students held an unusual demonstration in Leuven. Under the bored gaze of two policemen they protested against the separation of body and mind. Despite the very low turnout, the students split into two groups: one demonstrating for the idea and one against.

A 'silly season' story? Not at all! Although the roots of so-called 'dualism' extend a long way back in time – back to the works of Plato and Aristotle in fact – it remains relevant to this day. This philosophy, in which the mind is thought of as immaterial and the body as purely material, has had a major impact on how the current generation of intelligent systems are being designed. This is certainly true of robotics.

At its heart is the question of the way we distinguish between the robot as a body and artificial intelligence as a brain, between sensors and motors and between data and algorithms. Broadly speaking, an algorithm is an unambiguous description of a set of instructions to manipulate incoming data and thus perform a task. Its complexity can vary from calculating the largest common denominator of a fraction to working out the next move in a game of chess.

THE MECHANICAL TURK

The Mechanical Turk was a legendary mechanical chess machine invented by Wolfgang von Kempelen in 1770. The machine defeated Napoleon in a game of chess, but was later found to be a fraud: it turned out that it had a human operator hidden within it operating the pieces. We could call the mechanical machine the 'body' and the operator the 'mind'.

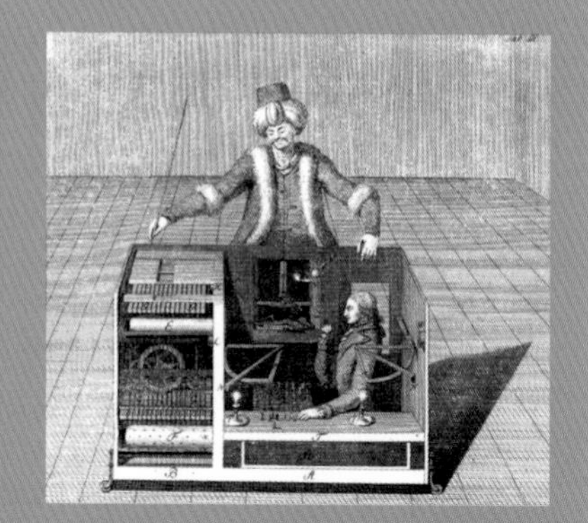

Today, we find this principle in crowdsourcing. Amazon's Mechanical Turk allows clearly defined tasks to be performed by people using computerised communication. Many believe that hybrid teams, automatically put together from human and AI agents, are the future of many business processes.

Embodiment

The philosophical dissatisfaction of some researchers with this way of thinking has led to a new movement in robotics called *embodied intelligence*. Rather than assuming that intelligence is created only by the mind, this approach proposes that it resides also in the body, and that there is interaction between the two. One example is human hearing: the fact that human beings have two ears a certain distance apart (a physical fact), allows us to estimate where a sound is coming from. This is possible because one ear receives the signal slightly before the other.

Embodied intelligence is based on three main pillars. First is the awareness that there is not one kind of intelligence, but that *the nature of intelligence depends on the physical computer*, the platform that executes the intelligence. Suppose we can describe intelligent behaviour using an algorithm: there are still several ways to perform calculations. For example, in addition to digital computers, there are also hydraulic, optical, chemical and biological systems that, in the most abstract sense, perform algorithms. The type of calculations and their efficiency vary greatly from one body to another. It is important to realise that the algorithm itself ultimately also changes the body, although often at a smaller scale – think of the neuronal structure of our brains, for example.

The second pillar is the *elimination of the distinction between 'input' and 'output'*, between sensors and motors. Embodiment assumes that there is an interaction between all entities and that input and output cannot be separated. For example, when human beings pick up an object we use not only our eyes and our knowledge of the world (for example, our expectation regarding the weight of the object), we also make intensive use of our sense of touch. Another illustration of the principle is the fact that there are several nerves running from the brain to the ear, suggesting that the brain plays at least as great a role in determining what we hear as the ear itself.

A third cornerstone is the recognition of the *importance of the ecological environment*. Just as intelligence cannot be separated from the body, it is also connected to the environment in which it manifests itself. We not only use the properties of our environment – as tuna harness the currents of the sea to accelerate – we also change our environment to help us perform cognitive tasks, for example by installing signs to help us navigate.

WIZARD OF OZ

Wizard of Oz experiments attempt to separate the influence of the body from that of the underlying intelligence. It is a technique often used in the early stages of the design of a robot and is a way of studying the interaction between man and robot. You could call it a kind of mock-up. The participant still communicates with the 'body', but the intelligent behaviour is simulated by the 'wizard' (often a human being).

One example is the Dragonbot social robot from MIT, which was specifically designed to help children learn. The child is not aware that the robot in this experiment is being controlled by a human being.

©Personal Robots Group, MIT Media Lab

There is no doubt that the division into body and mind offers advantages: it allows us to design highly complex systems and to work together efficiently. But this mode of thinking that dominates science and education – in which everything is divided into vertical silos, with mechanics separated from computer science, psychology from physiology, etc. – also has its dangers.

Artificial intelligence is a textbook example of interdisciplinarity, with AI designers needing to master at least three domains. First, the domain in which the intelligence applies: the rules of chess, linguistics, robotics or insurance, for example. Second, the mathematical and conceptual foundations of AI. And, finally, designers also need a sound knowledge of computer science and programming.

How does one measure (artificial) intelligence?

According to the above vision, intelligence is formed in co-evolution with the world and the body. Sensors provide experiences from which you can learn, but the level of sophistication of the manipulation with which you can change the world plays an equally important role. After all, if you can't use your body to experiment or to intervene in your environment, your options for learning are limited. This learning through interaction with the environment rather than on the basis of predefined data is one of the paradigm shifts that await us in AI, even though the ideas behind it have been around for decades. In other words: you use your body to express intelligence within your environment.

An obvious next step is thus to measure the quality of an intelligent system by examining its *impact on the world*. And this is, in fact, how systems are evaluated: how well did the algorithm succeed in performing a task?

THE BIRTH OF COMPUTERS

Ada Lovelace, daughter of the legendary poet Lord Byron, speculated at the beginning of the nineteenth century that computers might be capable of much more than performing calculations, putting her sharply at odds with her contemporaries. She envisaged these machines solving complex problems and even composing music. That's why many consider her to be the first computer programmer.

Even in the early years of artificial intelligence, the possibilities of computers were the subject of much philosophical debate. One of the key questions was 'can machines think?'. But thinking is a very vague concept and difficult to measure. That's why in 1950 Alan Turing, one of the fathers of modern computer science, proposed asking a different question, expressed in unambiguous terms. This is known as the Turing test and the question is as follows: is it possible for a human judge to distinguish between man and machine, solely on the basis of their behaviour?

This approach, in which we consider only the external characteristics, is sometimes referred to as *behaviourism*. We are all familiar with the iRobot, developed in the 1990s, which may not be intelligent, but does vacuum efficiently. When it comes to evaluating intelligence, we look only at the body and not at the mind. In other words, we disregard the question of whether or not the intelligent being has an 'inner world', saying that, according to this theory, each process (or thinking process) that produces the same results possesses the same degree of intelligence. This approach puts rote learning on an equal footing with higher insight.

Behaviourism is highly problematic because it does not take into account the quality of the thinking process. That's why we advocate *metacognitive machine learning*, in which the algorithms also reflect on their own learning process. How much thinking or calculating power does it take to perform a task? How much 'insight' does the solution have? How credible is the proposed model? How complex is it? Does the model include the symmetry and structure that is present in the problem definition?

AI OFFERS AN INSIGHT INTO HUMAN BEINGS

Artificial intelligence offers a unique opportunity for the pedagogical sciences to gain an insight into the mechanisms of learning. Unlike the workings of the human brain, computer simulations allow us to look inside the computer brain and carry out hypothetical experiments. A similar approach has already led to important insights in various scientific disciplines, such as language.

The image of our body in the world

All these questions are, of course, difficult to answer. They deserve further attention and are worthy of interdisciplinary research because they bring us right to the heart of AI. When an intelligent machine performs a task, it does so by observation and by interaction with its environment – through its body.

Incidentally, you shouldn't take the word 'body' literally. It could just as easily be a keyboard, a camera or data being read and written. A recent example is the iPhone's face authentication system. The device creates a model of the face using two cameras, resulting in a stereoscopic image and preventing the algorithm from being deceived by a flat photo. Such functionality is only possible by setting up two cameras, i.e. thanks to the body.

During that observation process, the machine creates a *world model*, i.e. an inner representation of its environment, and the more sophisticated this is, the more powerful is its intelligence. The possibilities of an intelligent system stand or fall by the quality of the model it uses, so it's hard to overestimate its importance.

A STREET MAP IS NOT A GPS

The term 'world model' refers to the information that an algorithm collects and stores about its environment. The requirements of such a model depend largely upon the task it is expected to perform. The more demanding and broadly applicable the task, the more sophisticated the world model needs to be.

For example, a street map, stored as a network of roads, is all you need to work out the shortest route between two points. For a GPS system, on the other hand, you also need to save the location of each street and node so that the current location can be connected to the network. Navigating a self-driving car also requires live images of the real-time environment.

The presentation of knowledge therefore remains a central theme in AI. Traditionally, there are two 'routes' for presenting knowledge: the *symbolic* and the *subsymbolic*. The first uses a top-down process to construct the world model, and its building blocks are so-called symbols and concepts that are comprehensible to humans, such as 'cat', 'tree' or 'street'. The power of these symbolic methods lies in *building new knowledge*. This is because logic-based algorithms can be used. They also offer *insight* into the reasoning process – not insignificant given the increasingly intensive interaction between humans and intelligent systems. The disadvantages, however, are that they require human intervention to express that knowledge, and are bad at coping with exceptions. For example, if you define a chair as a four-legged object, a three-legged stool will not be considered a chair. Of course, a definition can be extended, but in a realistic world there are so many exceptions and ambiguities that producing the correct definitions is far from trivial.

For this reason, so-called *subsymbolic* methods have been on the rise since the late 1980s. These use a bottom-up approach and start from the raw sensor input – nowadays often referred to as 'data' – which could be radar images, photos or sound waves. This highly dimensional input is linked to a class (fake/not fake) or an action to be taken by the intelligent agent. You could say that the system is *learning from examples instead of knowledge*. The advantage is that this information is much easier to gather, and the methods that build on it are less sensitive to exceptions. The disadvantage is that the complexity of the tasks that can be attempted is much lower. The creation of new knowledge is limited.

Language and art connect mind and body

Symbolic and subsymbolic methods can serve as models for learning through the mind or body respectively. Connecting symbolic concepts with physical artefacts is called grounding. Take gestures, for example. When someone waves their hands, we associate that gesture with the intention to attract attention. Connecting these two worlds remains one of the major challenges in the field of artificial intelligence.

In this context, it is therefore interesting to make the connection between language and creativity. Language is the ultimate tool for transferring thoughts, ideas and intentions – i.e. intangible things – from one mind to another. It is one of man's greatest and most ingenious achievements, and has brought us cooperation and dizzying advances through cultural transmission.

While with language we think we know the concrete meaning being conveyed, with art this is not the case. It is as if human spirits are communicating with each other without there being any external meaning behind their communication. The discussions above show that this does not by any means make it necessarily inferior. Food for thought.

Good?
WRONG?
RIGHT?
Should I...?

©VUB-Brubotics

DO ROBOT DESIGNERS HAVE PREJUDICES?

By Prof. Dr. Katleen Gabriels

'They must not forget for a moment that they are the guardians of their children and all those who live with them and are entrusted to their care.' This is an excerpt from an article published in the Flemish newspaper *De Standaard* on 27 March 1931. It's hard to imagine today, but when radio was first introduced in Catholic Flanders, this newspaper called upon the father of the house to take charge of the dial. This was to protect women and children from harm in their moral lives.

New technologies with far-reaching social impacts give rise to both hope and fear. In the case of the radio there was the fear of moral degeneration – a recurring fear with new technology – and, at the same time, the dream of bringing people closer together – a recurring hope. Upon the arrival of the world wide web, there were fears that we would lose the ability to distinguish between the virtual and the real, another constantly recurring theme known as *replacement fear*. At the same time, there was the hope that the Internet would give rise to an egalitarian society in which everyone's voice could be heard, regardless of looks, gender and ethnicity. Robots, too, give rise to both hope and fear. Hope, due to the many possibilities they open up, such as having them take over our heavy physical work. Fear of super-intelligent robots over which we will lose control.

The ethical debate about self-aware and super-intelligent robots is not the most pressing one: developments have not yet reached that stage. But there are plenty of other challenges within robot ethics that warrant our attention, such as the question of ethical design.

Ethical considerations can influence developments. All too often, we still think of ethics as something tacked on at the end after the technology has been developed. And we often hear that 'ethical issues shouldn't get in the way of progress'. That's why it's important to ask these questions right at the outset and throughout the design process.

Technology and ethics are not separate domains. Design is rarely neutral and norms and values often creep in, as do social prejudices and stereotypes. It's no different with robots.

Why design choices are socially and morally charged

In the Swedish fiction series *Real Humans*, hubots – a contraction of 'human' and 'robot' – live among 'real' people. Vera is a hubot that looks like an elderly lady and performs household tasks and caring duties. She wears an apron, likes to clean and is bossy and nagging; her (male) owner complains that she restricts his freedom. In short, Vera confirms all kinds of stereotypes about women, both in her appearance and in her behaviour. This example may be fictitious, but social stereotypes regularly creep into design. Why is a Ladyshave pink? It's difficult to think of this choice as 'neutral'.

Female sex robots and their pornographic appearance are criticised for similar reasons. Kathleen Richardson, robot researcher at De Montfort University, objects to these robots because they perpetuate the image of women as objects of lust and even as commodities.

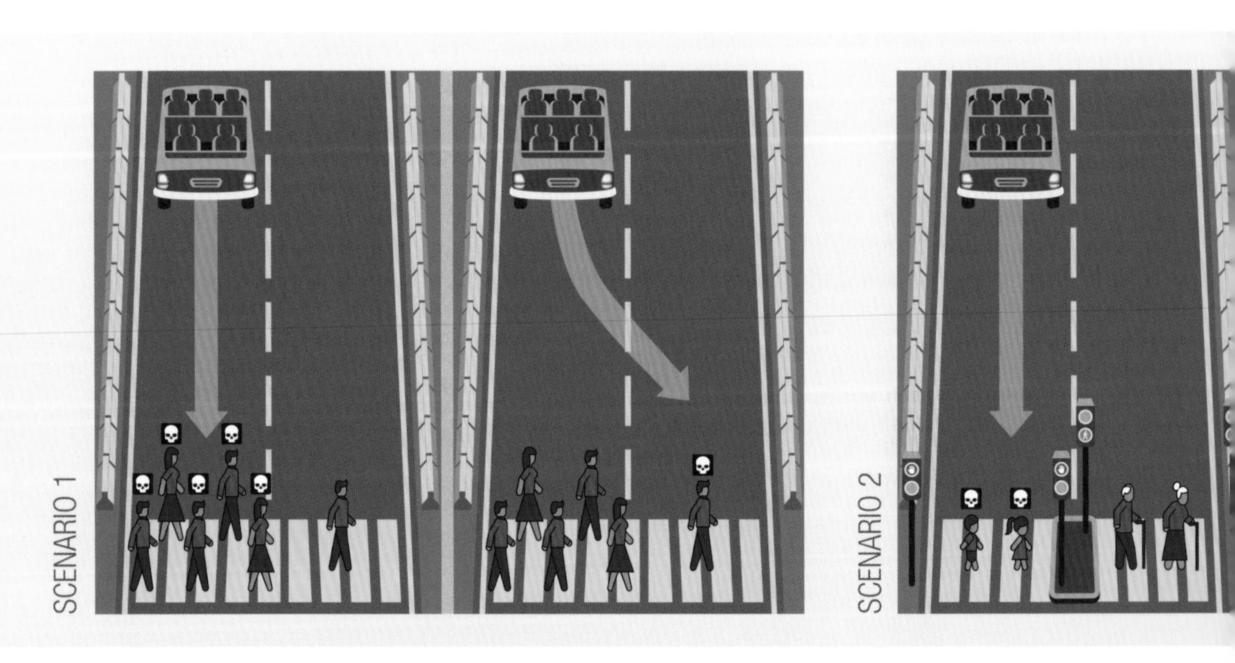

In her argument, she makes the link with prostitutes, stating that sex robots promote the further objectification and subjugation of women. In her 2015 'Campaign against sex robots' she started a debate about sex robots and social inequalities.

Even though Richardson's campaign has split the academic world, not least because her discourse is often hyperbolic, this is a debate that we need to have. Should we accept female sex robots or should they be banned? To what extent – and under what circumstances – are they desirable?

Robots don't design themselves: human beings are behind these choices. The history of technology is full of discriminatory design decisions at various levels. One example is the revolving door, which wheelchair users and people with walking sticks are unable to use. Debate is currently raging about algorithms that make discriminatory choices based on biased data. For example, a search for 'three black teenagers' on Google Images returns police mugshots, whereas the equivalent search for 'three white teenagers' results mainly in wholesome images of smiling teens.

In the MIT Moral Machine (moralmachine.mit.edu), researchers presented participants with moral dilemmas. Accidents caused by self-driving vehicles are likely to be unavoidable. Who should the car sacrifice: children, the elderly or the driver of the car? The research appeared in *Nature* and exposed cultural differences, including differences between Western and Eastern cultures. For example, the Japanese are more inclined to save the lives of older people than residents of Western European countries. However, the research is especially relevant to moral philosophers because it gives them an insight into the moral thinking of different cultures.

SCENARIO 3

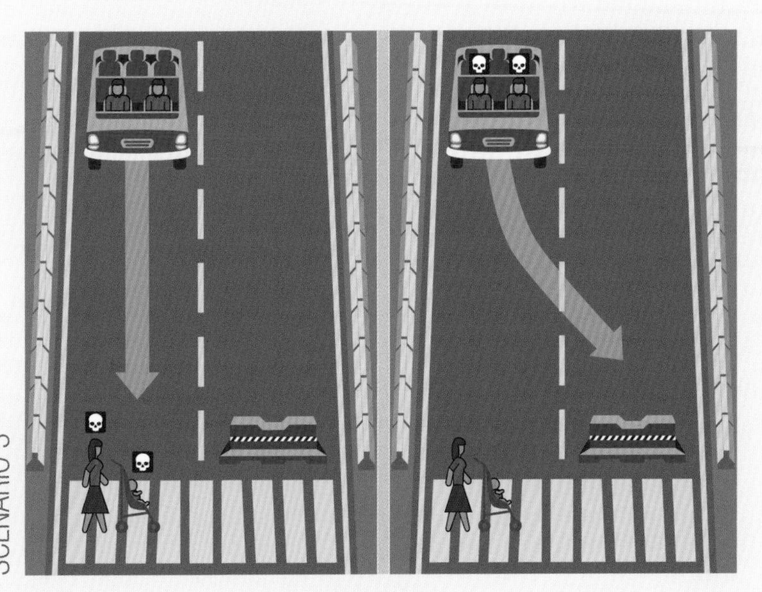

Better design

Technological design is at the heart of our society. The developers of the smartphone, for example, have undoubtedly helped to shape the way we live today. Tony Fadell, who was there at the start of the iPod and iPhone, has admitted that he regrets a number of design choices. When he designed the iPhone, he was unmarried and childless; today he can see the pressure the smartphone puts on his family life. When his teenage children are distracted by their phones, his wife says it's partly his fault. Fadell says they didn't take into account the smartphone in a family context, and he now calls this a design error – they took their childless lives as the norm. He now advocates a Hippocratic Oath for engineers and designers, similar to the one doctors take but imposing a requirement of ethical design. He says that such an oath would make those in involved in the design of technology more aware of their responsibility and the impact of certain choices.

The robot vacuum cleaner Roomba, which is equipped with sensors and a camera, caused quite a stir in 2017. The Roomba collects data so it can navigate within a house: a design choice that is necessary for the functionality of the device, but that also has implications for privacy. In 2017 it turned out that the CEO of the company behind Roomba wanted to sell that data to companies like Amazon.

Many robots – household robots and also care robots – collect personal data. For example, Lynx, a smart domestic robot equipped with Amazon's Alexa, includes a camera and built-in microphone. Lynx also has an alarm function that detects movement in the house when you are not at home and sends you a video clip. It's not always clear what happens to the data collected by such equipment. *Privacy by design*, where the data is destroyed after a certain time, is not always the desired design choice for a company: as the Roomba example makes clear, the stored data can also form the foundation of a revenue model.

Why design is inextricable from society

Technology and ethics are closely intertwined, as are technology and society: engineers don't design in a vacuum. However, there is no consensus on the extent to which the designer is responsible for these issues. At the ends of the spectrum lie two visions: separatism and technocracy.

The philosophy of separatism holds that engineers and designers are responsible only for technical decisions, with ethical questions being the domain of management, policy makers and users. As the previous paragraphs have shown, dividing the responsibility along such clearly drawn lines is neither self-evident nor realistic. The standards and preferences of designers may well creep into design, and their decisions often go beyond the purely technical and functional.

The German space engineer Wernher von Braun (1912-1977) was a notorious 'separatist'. Von Braun was involved in rocket development for the German army in the 1930s. He was also a member of the NSDAP – the Nazi Party – and joined the SS in 1940 because – by his own admission – he would otherwise have had to give up his dream of becoming a rocket developer. During the Second World War, Von Braun was involved in the development of the V2. This was the missile used to bomb London and other cities, but its highest death toll was during production, which was carried out by prisoners in concentration camps. Von Braun was aware of this. When he realised that the Germans were going to lose the war, he arranged for his team to work for the Americans. His work in America made the moon landing possible in 1969.

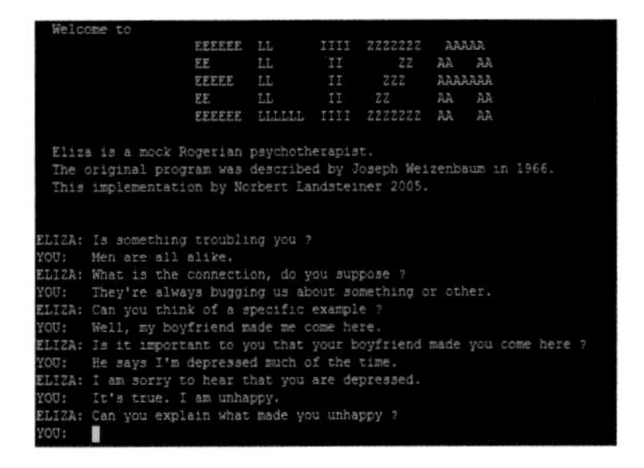

Von Braun did not consider the ethical and social implications of his work to be his domain. This is clearly a problematic attitude, and it's not only ethicists who point this out. One important voice in this field was Joseph Weizenbaum (1923-2008), an influential computer scientist and mathematician at the Massachusetts Institute of Technology (MIT) who, in the 1960s, developed the famous chatbot Eliza. His 1976 book *Computer Power and Human Reason* set out the human consequences of computers and AI. As a teenager, Weizenbaum fled with his parents to America from Nazi Germany. Throughout his life he continued to emphasise the intertwined nature of technology and society, and he was highly critical of scientists who worked for the Nazis and, later on, of the entangled relationship between MIT and the Pentagon.

MIT has developed and is developing a lot of technology for the government, often in the context of wars such as the Vietnam War. Weizenbaum rejected the separatist attitude towards responsibility, 'This attitude, to say, "I am a scientist, that is my discipline, and what happens with my research does not interest me. I am not a politician. Other people are responsible.' Separatism is a way to disengage yourself from moral responsibility and the need to think for yourself. This remains a hot topic – just think of the discussions about killer robots.

At the other end of the spectrum is technocracy. Under this approach, engineers and designers take all the decisions, including the ethical ones. But do they really have the necessary expertise and skills? Moreover, this is a paternalistic and undemocratic attitude.

Difficult ethical dilemmas

Today the development of autonomous, programmable technology, including robot technology, is in full swing. And one aspect of that programming is moral choices. The self-driving car is a good example. The autonomous private car was not developed all at once: rather, the process of automation has played out gradually over recent years to the point where we now have cars that park themselves and can brake or take bends autonomously. A fully automatic self-driving car raises interesting questions, because the car requires extensive programming. If an autonomous car is in danger of colliding with a bus full of schoolchildren, it can, in principle, be programmed to drive into a nearby tree to save the lives on the bus. But that may not be the choice that you would make if you were driving, because it would make it more likely that you would die – assuming that you were able to weigh up the options in this hypothetical scenario. The technology makes the decision for the driver. In principle, the consumer has a right to know how the car is programmed and which choices it will make. These choices and discussions are too important to be left to the technicians alone.

To raise social awareness about this issue, MIT has developed a website that presents such moral dilemmas involving a self-driving car visually. You can view various scenarios and design your own.

Neither separatism nor technocracy are desirable positions for robot design. Engineers and designers have a duty not to ignore moral questions during the design process (separatism), but also not to simply impose their moral choices (paternalism). Moreover, they need to look beyond pure functionalism (does the technology work properly?). Interesting methods exist that incorporate ethical questions from the very first reflections on the design, one of these being *value sensitive design*. These methods of technological design ensure that ethical values are upheld by embedding them in the design process and incorporating them into the technology. One example is *privacy by design*, which was mentioned earlier. The design process can also be democratised, for example by involving end users more closely and by having them give feedback as a form of co-creation. It is also important that engineers be given time to think about the possible use, abuse and ethical consequences of the technology they are developing. In order to familiarise them with these thinking methods and provide them with a 'toolkit' for dealing with such issues, technological ethics should be a standard subject for any future engineer, designer and computer scientist.

Conclusion

There is more to designing technology than simply its development and application, and the ethical issues involved extend far beyond functionality or standard safety questions. Due to its interdisciplinarity and highly complex nature, robot ethics is not an obvious discipline. Ethics is not an exact science: there is no formula you can apply and it's impossible to identify every potential consequence in advance.

Nevertheless, it is essential that we create a framework that incorporates ethical questions as a guideline as to what constitutes 'good' design that is in the public interest. It's not an easy matter, because who defines what 'good' is? And what does 'in the public interest' actually mean in any given circumstances?

But the good news is that the robot revolution is within our grasp. And we can achieve desirable results through cooperation, debate and a shared horizon.

'MURDER' BY ROBOTS

The media feed our fear of robots. A few years ago, newspapers reported that a robot had 'murdered' a worker in a German Volkswagen factory. But to be held responsible for murder, a 'person' must act freely and have awareness. So it's absurd to say that robots can 'murder' us. If you accidentally get run over by your own car because the handbrake wasn't on, we don't call this an 'attempted murder'. This type of reporting unnecessarily confuses the public about what robots are capable of.

In 2017, the British newspaper *The Telegraph* reported that a security robot had drowned itself in a fountain. Suicide requires self-awareness and free will.

But maybe in the long run this will be a good Turing test for robots: read *The myth of Sisyphus* by Albert Camus to your Roomba vacuum cleaner. This book opens with the words: 'There is but one truly serious philosophical problem and that is suicide. Deciding whether life is or is not worth living is to answer the fundamental question of philosophy.' And then quietly wait and see what happens.

AUTONOMY

A factory owner defending his workshop against Luddites attempting to destroy his mechanised looms between 1811 and 1816.

WHICH COUNTRY WILL SEIZE POWER OVER ROBOTS?

By Prof. Dr. Jonathan Holslag

There is a passage in the *Odyssey* about the women of Ithaka and their almost unbearable toil to grind the grain. 'Oh, Zeus, let this be the last day. My knees have been weakened by this exhausting labour.' As if answering the plea of Homer's toiling women, the Greek epigrammist Antipater wrote: 'Stop grinding, you women plodding at the grinding stones. Sleep yourselves out, even if the rooster crows, because Demeter has ordered the nymphs to take over your work, to throw themselves to the wheel, to set in motion the shaft that with its turning spokes will set the heavy millstones of Nisyros moving.' The water mills of Antipater took over the work of the women.

The men would continue to work with the flail for many centuries, until around 1830, when the threshing machine was invented. 'Work that used to take months can be done in a few weeks and the labour that is freed up can be devoted to more valuable things', was the jubilant cry. The English farmers themselves thought otherwise. They feared the bread was being taken from their mouths and used their flails to destroy the machines.

Technological progress, and in particular the ability to take over people's work or make it more productive, invariably gives rise to enthusiasm. Doing more with less effort: who could possibly object to that? But technological progress is invariably accompanied by uncertainty, and that's still the case today. Time and time again, optimism seems to turn out to be justified in the long run and the dreaded negative effects are not as bad as expected.

Just look at the threshing machine. Nowadays the farmer leaves it to get on with its job while he uses his tractor to take the harvest to the silo. Perhaps the tractor will soon be able to dispense with a driver altogether and artificial intelligence will take over the process of crop planning, so that all the English farmer has to do is invest in the right equipment. And there it is: the beauty of capitalism. Capital plus technology equals more productivity, greater prosperity and less work.

The possibilities in the field of automation, robotization, digitization and the application of artificial intelligence are endless, especially now that researchers are striving to surpass Moore's Law – which states that the power of computer chips doubles every two years – in all areas.

Difficult balancing act

We are evolving from productivity to almost limitless hyper-productivity. While there is no doubt about the potential of technological progress, it remains important to consider how it can best be harnessed in the service of human society. The advance of technology is a fascinating thing. It took the first humanoid people twelve million years to master the use of primitive tools. And it then took them another three million years to gradually free themselves from these tools and physical labour by the use of animals and the harnessing of external forces such as water, wind, steam and electricity. It now looks as if, in just a few decades' time, man will be able to hand over all his mental labour to external forces, thanks to the combination of robots and artificial intelligence.

This opens up opportunities, but it also means that societies are faced with a particularly difficult balancing act. They have no choice but to take part in the new race for robotization and artificial intelligence in order to strengthen their economic and military position. Those societies that fail to do so will become vulnerable, leaving them susceptible to the influence of other societies and jeopardising their safety and autonomy. It's a question of *the survival of the fittest*. At the same time, it is at least as important to think critically about how this technological progress can be transformed into opportunities to build a cohesive, conscious and happy society. After all, that is the condition for making power sustainable: the strength of a society can never be maintained in the long term without humanity. So societies have no choice but to join the technological race, but they should be wary of technological fetishism, because robots and computers alone cannot build a strong society.

American vs. Chinese giants

Companies such as Google in the USA and Tencent in China are gearing themselves up to become leaders in the field of robotization and automation. Will hyper-productivity – capital coupled with rapid technological progress – also lead to hyper-competitiveness? And what does this mean for prosperity and its distribution?

You could view the situation today as a new, multifaceted surge of creativity. Smart men and women working on tomorrow's technology over their latte macchiatos in hip start-ups. They are doing this in San Francisco Bay, but also in Nairobi, Xiamen, Hyderabad and Brussels. The Internet and the worldwide availability of investment capital have given rise to a finely-meshed global ecosystem in which everyone participates and everyone benefits. But reality paints a different picture. Although there are a great many small players, the economic battle for the future of robots and AI is mainly raging between the American capitalist giants like Google and the champions of Chinese state capitalism, such as Tencent and Alibaba.

Despite euphoric reports from China, the United States seems to be consolidating its leading economic position. The top fifty strongest robotics companies include thirty-two American and two Chinese companies. The United States is still ranked as very strong in AI, as well as in the production and innovation of computer chips.

Of course, we must take a critical approach to these rankings, especially as they were drawn up by American publications, but they do seem to paint quite a clear picture. This raises questions about the economic power of the American giants, with Google, Facebook and Amazon positioning themselves as the dominant gatekeepers of the digital market.

The American President, Donald Trump, has won many votes amongst the white working class in old, often declining, industrial states. He wants to 'make America great again' by bringing more production to America and prioritising American workers under the *America First* principle. Large companies such as Toyota are put in their place in 140 characters if they choose not to base their factories on American soil. Trump has also fired off his Twitter gun at the giant China, the factory of the world.

China on a shopping spree

For years, China has had one great asset: its cheap human labour. Even Donald Trump's red election caps were *Made in China*. But that is changing. Because of the one-child policy, which China has now been forced to abandon, the Chinese population is ageing rapidly. This means fewer workers, who are having to produce for more people. The Chinese work hard because they aspire to better wages and better working conditions – the Chinese middle class is currently the largest in the world. This is a threat to the advantage of cheap labour. Companies are already looking for alternatives in Vietnam and India.

But China wouldn't be China if it didn't have a big plan. And that plan is based on robots – lots of robots. The robot revolution, launched in 2014 by President Xi Jinping, is set to transform first China and then the rest of the world. Robots have become the cheap labour force, and their costs are still falling. They produce at 5 euros per hour, compared to 50 euros per hour for a German worker and 10 for a Chinese one.

China has a lot of catching up to do. The country has 68 robots for every 10,000 employees, compared to a global average of 69. The United States has 189 – approximately the same as Belgium and the Netherlands – whilst Korea and Japan have more than 300. But the Chinese figure is rising rapidly. Twenty seven percent of the robots produced worldwide end up in China. Moreover, the country is investing massively to allow it to develop this robot technology itself. As yet, Chinese robots are still of inferior quality, but the Chinese learn quickly. And they are going on a shopping spree. Chinese investors have recently taken ownership of the majority share of major German robot manufacturer Kuka, with Merkel still enduring criticism for allowing the deal to go ahead. And Belgian company Robojob, which manufactures robot installations for the metal industry, has also been partly taken over by the Chinese.

Europe lacks entrepreneurs

Europe has a great deal of knowledge in the field of robots and a strong market share of 32 percent, and its research programme Horizon2020, allows its researchers to collaborate effectively on a European and multidisciplinary basis. Moreover, the EU provides an important economic and social framework. But the associated entrepreneurship has yet to materialise. European start-ups are forced to mature quickly, with an emphasis on generating income and profit, rather than just growth, whereas American start-ups see growth as a priority: just look at innovative companies like Tesla, which has never made a profit. Even Facebook was happy to allow its user numbers to grow before investigating how to make a profit from them.

While European countries have many promising players, the poorer countries seem to be virtually absent from this field. This applies even to India, an economy that is highly regarded in ICT. Will robots and AI make it even harder for those countries to increase their prosperity?

Inequality is rising

We know all too well where European breakthroughs such as the threshing machine led in the 19th century. Whereas per capita economic output in Western Europe was twice that of Asia and Africa by 1800, by 1900 it was five times greater. Once Europe's industrial revolution had got underway, per capita production in Africa did not increase until 1950, and even after that, inequality remained enormous.

The latest technological upheavals have dramatically increased global income inequality, with technological power creating the capacity to perpetuate unequal economic partnerships and keep weak countries dependent. That is not exclusively a Western phenomenon, but the West was to first to achieve that dominance on a global scale. The International Labour Organization (ILO) warns that robotization and far-reaching digitization is a particular threat to low-wage workers in poor countries. In Cambodia, Indonesia and Vietnam, more than half of the workforce could be threatened with replacement by machinery. This trend is even more pronounced in regions with a fast-growing population, such as South Asia and Africa.

Thanks to robotization and AI, productivity will continue to increase, but there is a problem with the distribution of the profit from that productivity. Ultimately, countries only stand to gain if they are part of the innovative process, and this is not the case for many.

But the gap between winners and losers can grow within societies, too. Geographically, innovation tends to be increasingly concentrated around a few strong cities and regions, and one could argue that it is precisely here that the authorities have a role to play. They could regulate and redistribute, for example by introducing a basic income, but this is not what we see happening. Although governments do have the power to do this, they often have to compete to attract technology companies and tend to humour these companies. That was the reasoning of the European Parliament in its study advocating tax advantages for robots rather than tax redistribution. The social need for correction is great, but weak countries do not have the leverage to achieve it and strong countries do not seem inclined to use the leverage they have for fear of losing out in the economic race.

Europe will be swallowed up

This statement sums up Europe's uncomfortable position: its future is in the balance. It still has its strengths, but it is increasingly caught between the United States and China. Within Europe there is a huge discrepancy in terms of robotization and AI between Germany and the Netherlands on the one hand and the rest of the Member States on the other. If we are to put ourselves in a better position to make choices and to manage the social impact of robotization and AI in the future, Europe needs to be at the forefront of technological progress and be creating sufficient solutions for itself to avoid unbalanced dependency, thereby retaining the capacity to make autonomous societal choices.

Healthy market forces must be safeguarded. Small European players must be able to develop without being swallowed up by predominantly American near-oligopolists. Cross-border knowledge clusters will become essential, but so will investment levers. Europe invests less than a quarter of what the United States does in AI and the fight against industrial espionage.

To enable society to monitor the positive impact of robotization and AI in the long term, we as a society must also preserve our power and autonomy in that domain.

An AGV (automated guided vehicle) for the transport of electric motors and the pre-assembly of axles at Audi Brussels.
©Audi Bussels

WILL ROBOTS BE GIVEN A LICENSE TO KILL?

By Prof. Dr. Ir. Bram Vanderborght

A *licensed to kill* robot may be Q's ultimate technical gadget for James Bond, or perhaps M's ultimate secret agent, allowing her to dispense with the quirky secret agent altogether. But military robots are no longer confined to the realm of fiction. Even during the Second World War, the Germans were already using a caterpillar vehicle known as Goliath that they operated and detonated via a long cable.

The advantages of military robots speak for themselves. Robots show no emotion and can kill without pity or remorse, something only a minority of human soldiers can do. And the loss of a robot is less serious than the death of a soldier: after all, a robot does not leave a family behind.

As the technological revolution gathers pace, the expansion of this sector is in full swing and it is feared that robots themselves will be given the right to make the decision to kill. Many robotics and AI experts want to call a halt to this trend with a ban on killer robots.

A growing share in warfare

The importance of robotics for the American army is obvious. In 2000, the United States even passed a law governing this field, setting targets for at least one-third of operational attack aircraft in the US army to be unmanned by 2010, and one-third of attack vehicles by 2015. The figure for aircraft was reached without any problems, but the armed robotic vehicles are not yet in use on the battlefield because they are still too easy to overpower.

In 2005, five percent of air raids were unmanned and in 2016 more weapons were fired by drones than by manned aircraft in Afghanistan. Autonomous warships like the Sea Hunter, and submarines like the Echo Voyager, are also used to carry out tests. The US is expected to have more attack robots than soldiers by 2025. They want to strengthen their dominance in robotics in relation to China, Russia and the dozens of other countries engaged in such developments.

Money and body bags

There are two main reasons for the use of military robots: money and soldiers' lives. An F35 costs about 160 million euros, a smaller drone that soldiers can launch for themselves about 35,000 euros, and the famous Predator around 4 million euros. For the price of one F35 you can thus buy about 40 Predator drones. And Belgium wants to buy 34 F35s. The Pentagon estimates the operating costs of the F35 to be approximately 30,000 euros per hour, meaning that this is what it costs to use the Predator to guard the border with Mexico.

Public opinion also demands that war be conducted without victims: body bags are the great fear of any politician entering into a conflict. The First World War caused 16 million deaths, the Second World War 60 million, and the Vietnam War 315,000. Three hundred and eighty two American soldiers died during the first Gulf War and twelve during the war in Bosnia and Herzegovina. It is hoped that robots will mean even fewer lives will be lost.

Moreover, a soldier costs about 4 million euros from training to retirement. Military robot manufacturers like Foster-Miller and Endeavor Robotics (formerly owned by iRobot, known for its vacuuming robots) receive a lot of fan mail from the front about how many lives they have saved.

In aircraft, too, the pilot is the weak link, and a great deal of money needs to be invested in his safety. When a drone is lost, a new one is simply deployed. Drone pilots operate from comfortable seats in bunkers in the Nevada desert. After a day's work they go home to their families.

Today, body bags are an important argument for not going to war. It is therefore feared that governments will be quicker to resort to that option if fewer people and more machines are involved.

An MQ-1B Predator unmanned aircraft of the 361st Expeditionary Reconnaissance Squadron leaves Ali Base, Iraq, on 9 July, in support of Operation Iraqi Freedom.
©Tech. Sgt. Sabrina Johnson

Using robots creates the illusion that war is a risk-free endeavour. But if both parties are going to fight out their dispute using robots alone, they might just as well organise an online computer game: it's likely that a conventional attack will follow the robotic one. If only one of the parties in a conflict has robot technology, then the soldiers of that party are indeed freed from risk to a certain extent. After all, the chance of an operator in Nevada being killed in the battle is virtually non-existent. However, this may jeopardise the safety of citizens, as hostility can turn against them if soldiers are elusive. So waging war in this way is not without its risks.

On May 1, 2010, a bomb attack on Times Square in New York was foiled in the nick of time. In a video message, the Pakistan-born American bomber said that the attack was intended as 'revenge for the countless drone attacks in Pakistan' – drones are an object of hatred to locals there. So this is not really conducive to restoring the confidence of the local population at a time when this is of critical importance. We should therefore think carefully about the way we deploy robot technology.

New forms of war

During the Cold War, both the Americans and the Russians designed heavier, stronger, bigger and faster weapons, in quantities sufficient to destroy the entire world at the touch of a button. Later on, this technology allowed the Americans to end 'Operation Iraqi Freedom' after just 45 days. Standing in front of a *'Mission Accomplished'* banner on an aircraft carrier, President George Bush proudly declared that the war was over. Fifteen years and one billion euros later, Iraq is still a mess.

Fighter planes, aircraft carriers and other advanced war machines are of limited use against suicide bombers and improvised explosives along the roadside. This method of warfare has also spread to many other countries and regions.

Control room of an MQ-1 Predator unmanned aircraft.

©United States Air Force photo by Master Sergeant Steve Horton

Nowadays, war is often waged between unequal parties. Some call it asymmetric warfare, some call it terrorism. On the one hand is total technological supremacy, on the other an army with limited technology and a much lower budget. This was not the case during the Cold War, when both camps were striving to outdo each other.

The production of weaponry such as missile shields and nuclear weapons is hugely expensive and takes a lot of research. Iran and North Korea, for example, are struggling to develop an atomic bomb, even though this technology is more than seventy years old. Robots, however, are a different matter. Some programming skill and a little dexterity are all it takes to build an unmanned plane and to program it in such a way that it detonates when it hits the ground. IS has already built and used several of them. The drones are for sale in hobby shops: you can modify the software yourself and the internet is full of the relevant software modules and manuals. Such drones will not cause much material damage at first, but the psychological effect will be considerable.

For these asymmetric wars, the Americans want to evolve towards small, independent robots that go up against their prey as a team, like a pack of wolves that owes its strength and deadly efficiency to the fact that it works together. That discipline is called *swarm robotics*.

The robot makes the decision

It is still argued that robots are *human-in-the-loop models* - human beings are still involved. This strategy is certainly being applied today, but for how long? The pace of war is increasing all the time. Can a human being draw the right conclusions from the information they see on a screen, with a 'fire' button at their side? Often there are only moments to decide whether or not to press that red button. Soon, the instruction to soldiers 'don't think, the army does the thinking for you' will no longer apply to robots. They will be able to take decisions independently to fulfil a set task.

Opinions are split about this type of autonomous military robot. Some researchers believe that robots will never be able to distinguish between a civilian and a soldier, a prerequisite for compliance with the Geneva Convention, which forbids civilian casualties, or at least those out of proportion to military progress. The interpretation of video images, for example, does not yet make it possible to distinguish civilians from soldiers – unless you ask the enemy to wear a bright red helmet! Moreover, the difference between a soldier and a citizen is so vaguely defined that it often comes down to the common sense of the soldier. But common sense is the one thing that robots and computers don't have.

Experts also doubt whether robots can take into account the proportionality principle – i.e. that the benefits resulting from an attack must be greater than the damage inflicted – especially in complex and rapidly evolving situations.

Moreover, computer programs do not always work as predicted in real and often chaotic situations. After all, the time when robots did only what they were programmed to do is long gone. The programs that run within robots contain millions of lines of code, programmed by different teams of programmers, and no one individual fully understands how they work. This means that the behaviour of the robot is not completely predictable or particularly transparent. Moreover, such computer systems are also learning all the time using new techniques from artificial intelligence. Current AI techniques still have a great deal of trouble explaining and justifying decisions, which is essential when human lives are at stake.

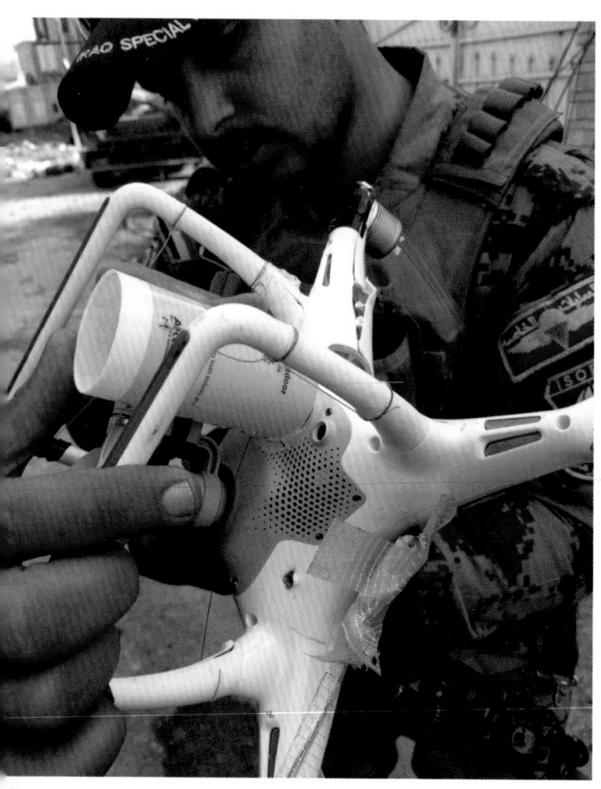

For example, an automatic warning system identified a Russian ballistic missile attack against the US. It turned out that the radars in Greenland had misinterpreted the reflection of a moonrise. A soldier decided not to fire, thereby preventing the start of World War Three. The passengers of Iran Air flight 655 were less fortunate. A warship owned by the US, which at the time was still supporting Iraqi dictator Saddam Hussein against Iran, interpreted it as a fighter plane and fired two deadly missiles. Human interpretation of decisions made by computer programs is still essential today.

Iraqi soldier of the Counter Terrorism Service in Mosul examines captured ISIS-drones. These are commercial DJI Phantom drones that were adapted to carry a 40 mm rifle grenade in the attached plastic tube and drop it above Iraqi troops.

©Mitch Utterback

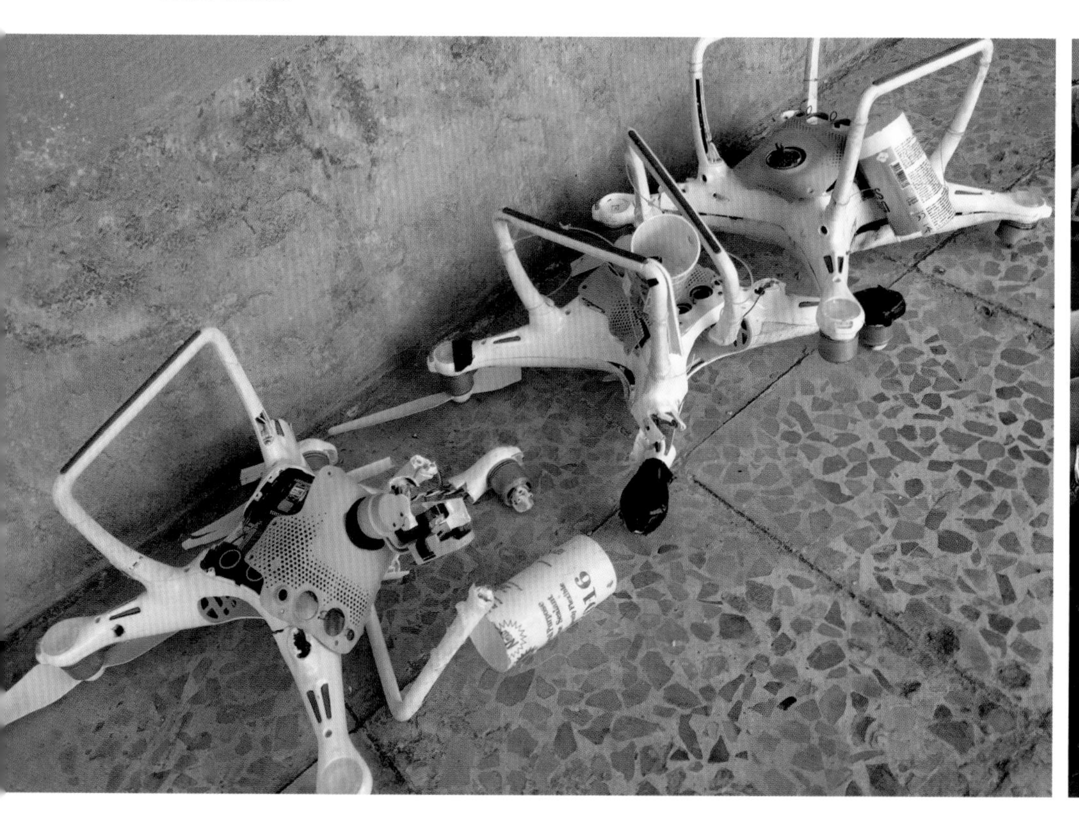

Robots don't feel anger or guilt

Other researchers think that the rapid development of artificial intelligence will give machines the edge over human beings. Computers are capable of processing huge amounts of data, and they are also more neutral and objective than people. A survey of soldiers who had served in Iraq showed that only 47% of the soldiers and 38% of the marines believed that non-soldiers should be treated with dignity and respect. Robot soldiers aren't affected by negative emotions such as anger or guilt. Moreover, a robot does not have the same need to defend itself as a soldier.

Belgium first country to ban killer robots

A broad movement is now underway against autonomous lethal weapons systems. April 2013 saw the start of the international campaign 'Stop Killer Robots', led by Human Rights Watch. And, in February 2014, The European Parliament adopted a resolution calling for a ban. However, marked divisions in the international community emerged at a meeting of the United Nations General Assembly's Disarmament and International Security Committee in 2015.

Nevertheless, in 2017 a number of international experts in robotics and AI, led by Elon Musk and Mustafa Suleyman (co-founder of DeepMind, an AI company bought by Alphabet-Google), called for a ban on killer robots. They wrote: 'Once developed, lethal autonomous weapons will permit armed conflict to be fought at a scale greater than ever, and at timescales faster than humans can comprehend. These can be weapons of terror, weapons that despots and terrorists use against innocent populations, and weapons hacked to behave in undesirable ways. We do not have long to act. Once this Pandora's box is opened, it will be hard to close.'

At the end of 2017, a number of Belgian professors also called on Belgian parliamentarians to support the resolution by Wouter De Vriendt (Groen) and Benoit Hellings (Ecolo) across party boundaries. They wrote: 'AI cannot take ethical considerations into account, because it has no compassion and no empathy. And those things are crucial when making decisions on matters of life and death.'

In 2018, the Parliamentary Committee on National Defence adopted a resolution prohibiting the production and use of fully autonomous weapons in the future. This makes Belgium the first country to adopt such a resolution. Of course, this is not enough on a global scale, but Belgium has played a pioneering role in the fight against anti-personnel mines and cluster munitions, and hopefully it can inspire other countries to follow its lead.

We need agreements

Perhaps the most important aspect of war is its psychological effect. After all, war is a dispute between people, and we should think hard before letting robots get in the middle of this. Robots on the battlefield certainly save lives, but we need to take a longer view. People used to wage war without machines, now people are waging war with machines: in the future will machines wage war without people? It is clear that an international debate is needed on military robots – and in particular on autonomous weapons systems – as well as a framework of rules covering this area. Without binding agreements, we are heading for immense suffering.

Many companies are making a lot of money from the new battlefield robots and, in that race, robot ethics is seen largely as an obstacle to be overcome.

As in so many fields, bringing together the complementary strengths of man and machine is what will produce the best results. Although we would like nothing more than for the ideals of peace to be realised, that's a long way from the international reality. A modern military with intelligent use of cooperating robots, adapted to the new type of warfare, is essential.

©Cpl. Levi Schultz

Lola, a self-propelled robot from the 1990 VUB AI Lab, navigates between obstacles and learns from experience.

HOW DO WE STAY IN CONTROL OF BILLIONS OF AI AGENTS?

By Prof. Dr. Luc Steels and Prof. Dr. Katrien Beuls

The aim of research into artificial intelligence is to build smart 'agents' using computer programs. In the past, we used to talk mainly of knowledge systems, now we call them bots or AI agents – the term we are going to use here. The word agent can be found in terms such as travel agent, secret agent, insurance agent and agency. An agent is someone who has the capacity to make rational decisions independently and, if necessary, to execute these decisions, for example to come up with the best plan for traveling from Brussels to Venice and then buy tickets. Or to take out fire insurance that takes into account the risks and the client's financial situation. An agent is also often a representative of a person who, for whatever reason, has delegated a certain decision because they cannot be present or do not have the necessary knowledge to act on their own behalf. Because the world is changing and new challenges keep cropping up, agents also need the ability to acquire new knowledge and make better decisions.

Just as you only have a travel agent buy tickets if you agree to the price, an AI agent does not have to work entirely independently. Thus, most AI agents are tools to help people perform their activities, and give advice without taking further action. Some are very simple. In your computer, for example, there is an agent that monitors energy consumption and puts your computer into sleep mode before it runs out of power and loses your data. But agents can also be immensely complicated, such as the one that defeated the world chess champion, or an agent who knows and applies the law to resolve a dispute about inheritance rights.

First and foremost, a robot has a number of mechanical aspects: a body, motors that make its limbs move, senses, a battery. If we also want a robot to be able to make decisions and interact with humans through language, then it needs to have an AI agent. We call this an *embodied agent.* An embodied agent must be able to seamlessly link mechanical aspects to mental ones: this is now known as the *grounding* problem and has proven very difficult to solve. Most agents are not embodied and to emphasize that, they are sometimes called 'softbots'. Moreover, most robots do not have an agent in them. They are pure machines that perform only routine operations – albeit automatically.

They are everywhere

There is little doubt that AI agents can be useful. An agent that controls and optimises electricity consumption in private homes with the user's consent can significantly reduce that person's energy bill. And the AI agents behind search engines like that of Google also help us to find the right information about search terms or questions on a daily basis. We dream of self-driving cars, but that is only possible if, as well as the mechanics of the car, there is also an agent monitoring the environment, following a plan and deciding when to turn off or brake.

We may not know it, but we are already surrounded by agents. A computer or smartphone is full of AI agents, and they are behind many of the tasks performed in companies or authorities. They check for the improper use of your credit card, analyse photos of your skin to see if cancer is developing and schedule trains. Sixty million bots are active on Facebook, and bots send 15 percent of all messages on Twitter.

However, it has become clear that AI agents are not always beneficial or effective. We are horrified by the idea of agents secretly gathering information about us by snooping around in our computer, or sneakily manipulating the price we have to pay for a plane ticket because we have already shown an interest in a particular flight. You know it's gone too far when agents fake the news you get to see or follow you around with advertisements for products you don't want.

The Swarm Bots from the lab of ULB's Prof Marco Dorigo are simple robots that are not centrally controlled, but can still work together to solve complex problems. This allows the robots to join together to overcome obstacles. This approach is inspired by the behaviour of ants, for example.

©Marco Dorigo, IRIDIA, ULB

They are not people

However fantastic AI technology may sometimes be, there are also some serious flies in the ointment. The idea that AI is already comparable to or can even outstrip human intelligence is a dangerous misunderstanding. This lie is mainly put out by people who want to draw attention to themselves, or who have commercial interests to push.

Firstly, there are many processes relevant to intelligence that we do not yet understand, or at least not in such a way that we can build computer systems to mimic them. There is still a lot of mystery surrounding the way in which we understand or learn language and form new sentences. Human memory is incredibly efficient because it can selectively store facts and magically retrieve those that are relevant to a particular problem out of the millions or even billions in our heads. We humans can come up with totally new ideas, solve a problem by analogy with another problem, learn the motor skills needed to play music scores with feeling, recognize faces at lightning speed and much, much more. We can also feel emotions when we listen to music and experience empathy for others, and we have awareness and self-awareness. These phenomena have not been investigated in detail from the point of view of AI, so there is more than enough fundamental research to be done in this field to keep us busy for decades to come.

Secondly, there are limits to way in which we can build AI agents. In fact, there are only two possibilities: either we have to build human knowledge into an AI agent ourselves, or have the agent gather knowledge via statistical learning algorithms. In the former case, we must first analyse the problem. This can be done by introspection, by systematically questioning domain experts or by reading and systematising scientific insights from articles, books and encyclopaedias. This is not only very time-consuming work, but human knowledge is also limited. Experts often disagree and are frequently unable to explain what they know properly. And a lot of knowledge is passed on orally, rather than being found in books.

Substantial efforts have already been made to create AI agents that can read texts on the internet for themselves and thus collect and store information in gigantic knowledge networks. IBM's demonstration of Watson, an AI agent that was able to beat human beings in a TV game in which common sense plays a major role, is based on that technology. Search engines also use this technique to better understand a user's question and to shape the answer more clearly. Google, for example, has gathered a knowledge network containing 70 billion facts, partly by 'reading' the articles of the online encyclopaedia Wikipedia and pouring them into formal structures that can then be used by language and reasoning systems.

The second way to build agents is based upon statistical learning algorithms and is called *machine learning.* A statistical learning algorithm uses a large number of examples as input and looks for patterns to make predictions or to recognise situations. The more examples there are, the better. Some learning algorithms draw inspiration from the neural networks found in the brain. But AI doesn't really attempt to imitate the brain in a realistic way, partly because we still don't know enough about its workings. Airplanes were not invented by imitating birds but by systematically tackling the problems that need to be solved to achieve artificial flying.

Black swan

Statistical learning algorithms have been around for a long time. Most of them used to be known simply as statistical methods, without any mention of AI. What's new is that these methods no longer have to be calculated manually by human beings, but are programmed within computers. Moreover, we now have enormous quantities of data, and the capacity to store it. Computers have also become much cheaper and more powerful, so algorithms that were once too time-consuming to be useful now produce a result within a reasonable period of time. Once a decision network has been learned, it can be executed efficiently, even with little computing power.

The recent attention and enthusiasm for AI is largely based on the large-scale application of these statistical learning algorithms. But here, too, there are serious limitations. First of all, you need enough data, otherwise statistical generalizations are not very reliable. If you want to be able to make a decision about a group of people – for example how high the risk of cancer is in smokers – you need data on enough cases to calculate the probability with reasonable certainty. Secondly, the data must also be reliable and reflect the natural variation in the target group.

And, even then, things can go wrong. If we give examples of 999 white swans and 1 black one, the chance of a black swan is only 1 in 1000 and a statistical algorithm will predict, with great confidence, that the next swan we will see will be white. An AI agent following the same reasoning will look for white moving objects when looking for a swan in an image. But, of course, there's always a chance of a black swan. So a prediction may be wrong, or a pattern may not be recognised because it has been rejected as too improbable.

There are statistical learning algorithms – *deep learning* being one of the most prominent – that can work with multiple layers. This means that they first learn basic categories (for example, colour, shape, weight), then more complex categories that group together a number of basic categories (for example, something of a certain colour and shape such as a white teddy bear) and then combine those more complex categories (for example, by taking into account

SONY

The Talking Heads experiment from the VUB AI Lab in 1999 comprises two AI agents, who in turn consist of a camera and a computer. The cameras are aimed at a board full of coloured geometric shapes and the computers communicate about what they can see on the board. In a Language Game, the AI agents developed their own language during this communication.

©Marleen Wynants

extra properties, such as a white teddy bear that falls to the ground). This gives rise to more powerful AI systems because stronger generalizations are possible. But there is another problem. The categories that such learning algorithms find are almost never the ones we humans use, which makes it hard to understand how an agent is making a decision. This also makes it more difficult to question decisions – a serious problem for AI agents that deal with people, such as legal knowledge systems. Nobody wants to be convicted without knowing why or without being able to argue their case.

The dangers

These fundamental limitations of artificial intelligence don't mean that we should throw the baby out with the bathwater. When a new technology appears, there is always a tendency to both underestimate and overestimate it. Underestimate, because it's difficult to imagine its impact and because people will always do new and unforeseen things with technology. Overestimate, because the limitations of a new technology are often not recognised.

This is very clear in the case of AI. For a long time it was thought that AI research would lead nowhere, and no serious investments were made in it, particularly in Europe. This goes a long way towards explaining why the industrial application of AI in Europe has fallen so far behind that in America. But the opposite is now happening. People think AI is approaching the level of superintelligence and is therefore dangerous, perhaps even a danger to humanity itself.

The truth, as is often the case, lies somewhere in the middle. Today there is much more to AI than most people think, and its use goes far beyond what is evident at first sight. But a demonstration and a real-world application are two very different things, which means that some research results are misleading. The main danger of AI lies in the possibility that it will be misused for financial gain or that people will fail to develop their knowledge and skills sufficiently and end up being less smart than the machine. For example, if you always use digital aids to navigate a city, you will never really get to know its geography.

The example of nature

Let us now take a look into the future. There is little doubt that the number of AI agents in our world is steadily increasing. This evolution is taking place alongside population growth and the complexities and problems of overpopulation. We are gradually ending up in a situation where billions of agents operate in the Cloud, almost imperceptibly playing a role in nearly all human activities.

But that is giving rise to problems we can't yet see. Who will develop all these agents? And if they learn for themselves, will we be able to stay in control of them? Won't there be situations in which some of those agents – either acting on their own or under the control of unscrupulous individuals – cause harm to all of us? Recent phenomena in digital media show that manipulation by unprincipled individuals is already commonplace.

Construction robots that collaborate to create complex structures out of building blocks without any central control.

©Marco Dorigo, IRIDIA, ULB

Thanks to technological developments, researchers at the VUB AI Lab no longer have to get to work with screw-drivers and soldering irons before they can test their experiments. They can now do this using the NAO robot, for example, which can be used for demos or research. The setting up of Language Games is one possible application.

One example of this is the scandal surrounding Cambridge Analytica, a company that illicitly collected data and used it to manipulate elections. And how will all these agents work together? It will no longer be possible to set up a central control organism to manage everything as the number of agents and the variation in their behaviour will be too large for this approach.

We believe that models from biology offer a solution. Living nature is made up of billions of organisms that all act autonomously, adapt, organise themselves and evolve over time. Since Darwin's time almost two centuries ago, biologists have been studying the mechanisms behind the remarkable robustness and complexity of living beings. Nature has succeeded in developing a 'technology' that still surpasses that created by human beings. The human brain is one of the most beautiful examples of this.

Another is the amazing processes of genetics: information is stored in DNA and this information is transformed and expressed to build and maintain an organism. We are still discovering, on a daily basis, new and miraculous ways in which organisms survive, sometimes in extreme circumstances. Bees and ant colonies show that it is possible for large numbers of organisms to work together without central control. Can we use biologists' insights into the basic mechanisms behind these peculiar properties to allow the massive quantities of future agents to grow and evolve?

Robots develop language

This line of thought is being investigated in a branch of AI called 'artificial life' in which the AI lab of the Vrije Universiteit Brussel (VUB) has long played an important role. Our research is mainly focused on the creation, further development and adaptation of language, and we use biological principles to create these phenomena in a digital medium. Just as human beings have developed a range of languages and dialects – and are still creating specialised sub-languages for new domains – artificial agents must somehow be able to find new conventions for communication and share these conventions with each other.

We are investigating this based upon experiments with autonomous agents that use language to communicate with each other about objects in their common world. In fact, they play a kind of language game. What's different about today's AI is that the language is not fixed, but is created and evolves based upon the activities of the agents themselves. It all seems to work very well. It is an example of the 'new' AI, based not on direct programming nor on learning an existing corpus, but on biological principles about how complexity can arise in autonomous beings.

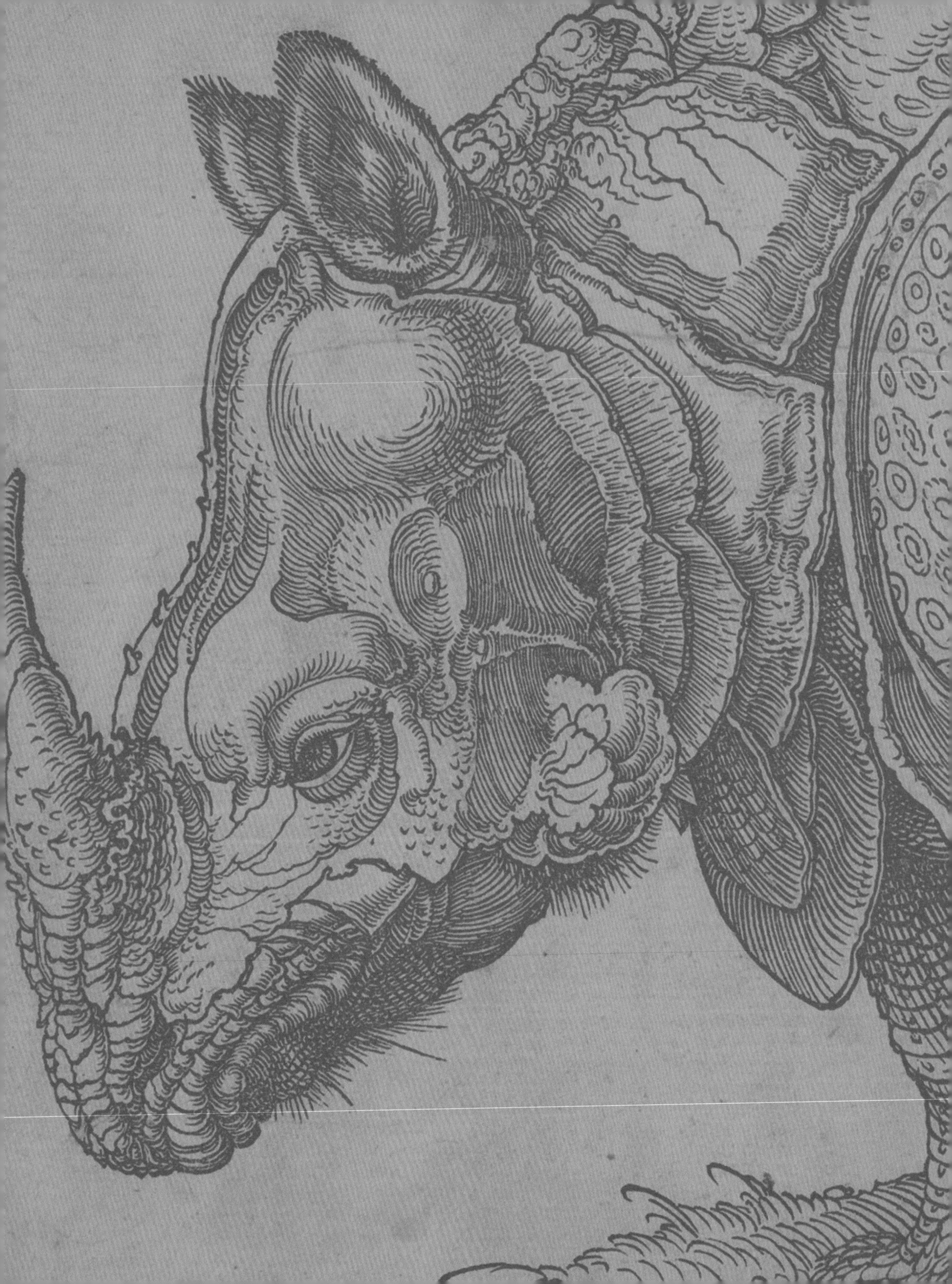

WILL WE SOON BE SENDING OUR CHILDREN TO THEIR SPORTS CLUB ALONE IN A SELF-DRIVING CAR?

By Prof. Dr. Cathy Macharis and Prof. Dr. Lieselot Vanhaverbeke

In 1515, artist Albrecht Dürer made a woodcut of a rhino, despite never having seen this beast with his own eyes. Based on descriptions, and a sketch by an unknown artist, he created this commendable work of art, which was subsequently copied many times.

This metaphor serves as a beautiful illustration of the way we talk about autonomous vehicles and robotics. As yet, none of us have seen for ourselves what they could look like in the real world, and what reaction they will evince from people, vehicles and the environment. Basing ourselves upon partial studies and pilot projects, we attempt to gain an insight into the way in which a mobility system involving self-driving vehicles could work.

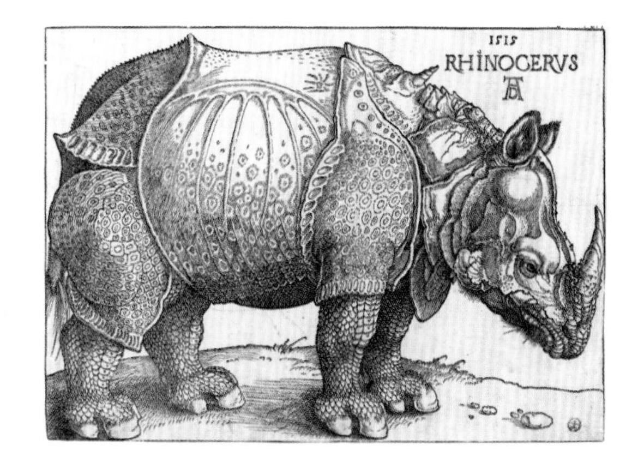

What is a self-driving car?

So, how does a self-driving car actually work? In the most basic terms it is a vehicle that can drive without human intervention. The current prototypes usually look like cars as we all know them, but are loaded with technical gadgets: cameras, radar, lidar (an advanced version of radar based upon lasers), GPS and movement trackers. Based on the information collected by all these sensors, the vehicle's software uses state-of-the-art algorithms to decide how it should drive in accordance with the rules of the road. This is what is known as an autonomous vehicle.

If the car is also connected to other cars – directly or via a control centre – and even to the road infrastructure itself, this is known as a connected autonomous vehicle. The advantage of connected cars is that they can take into account the local conditions – for example roadworks – and also the planned routes of other cars when plotting a route. This allows the best route to be calculated for each car on the road, according to the predicted available road capacity. Autonomous connected buses can even incorporate calls from waiting passengers into their dynamic timetable and flexible routing during their journey, thereby greatly improving their service.

Everyone is experimenting

A great deal of experimentation is currently being carried out with vehicles equipped with these technologies, both on abandoned industrial estates and right in the centre of our towns and cities. The level of automation varies. For example, a Tesla self-driving car is already capable of driving fully automatically on major roads and motorways, but still requires human supervision and intervention in difficult situations. 'Level 5' self-driving vehicles, on the other hand, operate completely autonomously in all circumstances.

In Belgium, Vias and FPS Mobility are testing self-driving shuttles in regions such as Waterloo and Han. Another example is the project of the Flemish bus company De Lijn involving a fully-fledged autonomous vehicle (i.e. one with no human intervention at all) driving in mixed traffic. Technical validity tests are currently underway on a route around Zaventem airport. Passengers will be able to take the self-driving bus in 2021.

A truck platooning test – in which the first truck has a human driver and the others follow automatically – has already been carried out. In addition, a self-driving Vias shuttle has been tested on the Francorchamps circuit and a self-driving delivery bus has been demonstrated in Mechelen, as has a self-driving car on a private track in Leuven. And there are many new projects in the pipeline. For example, there will soon be an autonomous shuttle on the campus of the University Hospital of Brussels.

In the United States, self-driving vehicles have been making their way in amongst the normal traffic for a number of years now, albeit with a so-called *safety pilot* behind the wheel, who monitors the movements of the vehicle and can intervene if necessary. Uber, Cruise and Waymo are the best-known examples, but Audi, BMW and other car manufacturers are also working in this field. It is also worth noting that these projects often involve collaborations between a technology company and a traditional car manufacturer: Uber has struck a deal with Volvo, Cruise has been acquired by GM, and Waymo belongs to the Google group Alphabet and works with cars from Fiat Chrysler. This shows that the sector is in a state of flux. And, thus, also that our mobility may well look very different in the future.

©2getthere - De Lijn - Bussels Airport

Will we want to take our hands off the wheel?

Our current mobility system allows us to get about, but there's a downside. Worldwide, an estimated 1.25 million people die every year on the roads, 25,300 of these in the EU and 620 in Belgium. Then there are the traffic jams. And vehicles with an internal combustion engine produce emissions, which gives rise to air pollution. Does self-driving technology offer a solution to these mobility challenges?

In any case, the first question must be: will we want to take our hands off the wheel? That may be quite scary at first, but people who already travel as passengers on a regular basis will probably take it in their stride. Our experiments indicate similar levels of adaptability to those for the electric car: once test subjects had the chance to ride around in it, responses were usually positive.

But what will these vehicles look like in the future and what will passengers think of the ride? It turns out that a lot of people experience motion sickness in self-driving cars if they are not looking through the window. For those currently suffering from car sickness, research into the technology of self-driving cars is a good thing. Car manufacturers are investing heavily in studies into methods for combatting nausea, ranging from advanced shock absorbers (Clearmotion), internal projection screens and vibrating seats (Uber) to special glasses (University of Michigan).

How will this affect the number of cars on the road?

A lot of research is currently underway into the effect of autonomous vehicles on travel behaviour: will it increase or decrease the number of miles people cover? It looks like it will up the mileage. That's only logical, as the time it takes to get from one place to another will become less of a factor. People won't mind sitting in their cars so much if they can get some work done, or play with their smartphones. And more people will be able to travel by car. Young people without a driving licence, people who are unable to drive due to a disability and elderly people who have decided their driving days are over will all swell the numbers. Autonomous mobility means greater freedom for these population groups.

Most studies therefore assume an increase in mileage of between 40 and 80 percent. But we don't know for sure – we haven't seen the rhino yet. A neat (albeit expensive) experiment into the effect on travel behaviour was conducted in the US, in which thirteen families were given a car and driver day and night to see whether this increased their car usage. And, guess what? Their mileage increased by 83 percent, especially in the evening, and often to give other people lifts. And, yes, the children were only being driven to the sports club.

BMW's self-driving 5 Series during a test drive on a motorway.
©BMW

So self-driving cars are not going to do away with traffic jams. But it does seem possible that this technology will allow us to make more efficient use of shared mobility solutions. Shared self-driving taxis can get people to their destinations, sometimes in combination with public transport. A simulation study by the International Transport Forum for the Lisbon region in Portugal showed that the congestion problem could be solved if we shared vehicles on a large scale. Emissions would also be cut by half.

This study assumed that it would be easy to 'order' a journey, for example using an app, and that a vehicle was then quick to arrive at requested location – like an Uber but without the driver. This is very convenient for users:journeys can often be completed without changing trains, or are oriented around good connections to public transport. It cuts the costs by half, which helps achieve greater equality for all mobility users. In such a subscenario, the purchase price of a vehicle ceases to be a factor and people pay for transport based on their usage.

A similar study for Brussels indicated that only 83,000 shared cars would be needed instead of 730,000 private cars for all traffic within, and going into and out of, the city. The existing system of *free-floating* sharing cars – without a fixed location – is already moving us in that direction.

There will be fewer accidents

In any event, we can expect road safety to increase. Since 72 percent of all accidents are attributable to human behaviour, we can assume that smart cars won't cause as many. But that's the nub of the problem. Because public tolerance is much lower for accidents in which autonomous vehicles play a role: a case in point being the accident on 18 March 2018 involving an Uber car.

In the field of road safety, it is generally assumed that an autonomous vehicle will be accepted by the public as safe if only one incident is reported per 100 million hours of autonomous driving. Another question is whether you set up the algorithm to save the life of a girl crossing the road or the person in the car? This is why car manufacturers are very reluctant to launch autonomous vehicles onto the market. Road safety first needs to be absolutely spot on, or as good as. It will be some time before all possible situations have been properly mapped out, and the system may never be truly watertight.

However, many people are hard at work on this issue. In 2017, Waymo reported that it had covered 525,000 kilometres in tests of the technology in California, and a total of 12 million kilometres on the road since 2009.

This explains why the self-driving shuttles that have been tried out here and there stay well below their top speed, driving around at under 30 kilometres per hour. This allows them to minimise the damage in the event of a collision.

Are we nearly there yet?

So, when will be all be able to take a trip in a self-driving car? That's more difficult to predict. The major car manufacturers have announced, to great fanfare, that these vehicles will be on the market in 2021. Academics working from a systems perspective cite 2035 to 2050 as the time horizon for an autonomous mobility system.

The blind Steve Mahan about to test drive a self-driving car from Waymo, a company founded by Google. The car no longer has a steering wheel.

©Waymo

Impact on public space

Another subject for speculation is the effect on town planning. It is expected that autonomous vehicles will reduce the space needed for cars in our towns and cities. It will be possible to use the road infrastructure more efficiently - for example, four lanes for cars to drive out of town in the evening rush hour and just one in the mornings. Cars will also be able to drive closer together, again saving space.

In addition, far fewer parking spaces will be needed. At the moment, cars are stationary for 95 percent of the time. This figure will be much lower for autonomous shared cars.

But that's the big question: will self-driving cars also be shared cars? If not, aren't we just making the traffic problem worse? In the longer term, people may also live further away from town centres and rely on autonomous vehicles to bring them in.

But towns could also be organised differently. Shared autonomous vehicles would mean a saving in land for public parking spaces of no less than 95 percent. Since the number of parking spaces in Brussels (around 265,000) would cover the distance from Brussels to Rome, or around 1,500 kilometres, it would be quite nice to get that space back!

Moreover, a system using only shared autonomous taxis and taxi buses would require only 3 percent of the number of vehicles currently on the road. Each vehicle would cover up to ten times more distance, but it is expected that 37 percent less distance would be driven in total, even at peak times.

A SINGLE FATAL ACCIDENT

The first fatal accident involving a self-driving car took place on March 18, 2018 in Tempe, Arizona. It is being blamed on Uber. There had already been a few crashes in which the driver of a car on automatic pilot had been killed, but this was the first to kill a third party, a pedestrian pushing a bicycle across the road. The self-driving vehicle failed to brake and the safety pilot was distracted just before the accident and didn't have time to intervene.

How do we turn this into a positive story?

These days we are hearing lots of wild stories about our rhino, the autonomous car. We may already have seen a few rare specimens, but we can only speculate about an entire population with the potential to have a massive influence upon our mobility behaviour. What is clear, however, is that the self-driving car will change not only our travel behaviour but our entire way of life.

If we want the robotization of the transport sector to evolve into a positive change that can help us face our current mobility challenges for the benefit of all, now is the time to develop policy measures. Such measures can help us to put in place the right context so that we can prepare our entire mobility system for this kind of positive evolution – or even revolution. We need to encourage car sharing and shared transport, create clear frameworks for mobility providers, stimulate innovation and new business models and, at the same time, protect users.

Europe failed to respond to Uber until it had already put down roots in our cities, and this mobility service is now so well established that it is impossible to imagine the city without it. Imagine a technology giant in Brussels with a fleet of self-driving cars that respond to transport needs *on-demand* and in a user-friendly fashion. Failure on the part of the city or region to prepare the transport sector, infrastructure, citizens, and so on, for this development will reduce the chances of a positive transformation. It is crucial that we devise and implement a vision proactively before the rhinos actually arrive.

BMW is working on a R1200 GS self-driving engine. The aim of this prototype is to gather knowledge about driving dynamics and the detection of dangerous situations to assist the driver.

©BMW

WORK

Robots at Audi Brussels assemble the battery pack of the electric Audi e-tron.

©Audi Brussels

WILL ROBOTS TAKE OUR JOBS?

By Prof. Dr. Luc Hens

Watch out! The rise of ever smarter and more convenient robots is set to cause massive unemployment and do away with your job. Titles like *The advance of robots: how technology will make many jobs disappear* don't beat about the bush: they speak clearly of the fear that automation will cost us our livelihoods. This book by Martin Ford was awarded the prize for business book of the year by the *Financial Times* newspaper and the consultancy firm McKinsey. MIT economists Erik Brynjolfsson and Andrew McAfee also warn of the effects on employment, and on the gap between rich and poor, in their bestseller *The Second Machine Era: Work, Progress, and Prosperity in a Time of Brilliant Technologies.*

The fear that automation will give rise to job losses is not new. Between 1811 and 1815, farmers and artisans in England destroyed mechanised looms because their cottage industry was threatened by the emerging textile factories, and the 1920s and 1960s also saw waves of fear regarding the effects of automation. In the 1980s, English printers and typesetters went on strike for months because their jobs were disappearing due to the rise of offset printing presses.

The shrinking middle class

The shifts in the labour market are real and not limited to the United States. But they are also more complex than might appear at first sight.

To understand how automation affects employment, it is best to consider work as a package of tasks, some of which are now being taken over by machines. Routine tasks are usually easy to automate, whilst others are more difficult. This may be because they require complicated manual operations – just think of the sensory-motor skills you use to fold a towel – or because they call for intuition, creativity and persuasiveness, or the ability to solve problems. According to economists Deirdre McCloskey and Arjo Klamer, abstract cognitive activities of this kind represent up to a quarter of the gross domestic product.

For example, an accountant has to tot up figures in tables: routine work that is easily taken over by a computer spreadsheet. But working out what these financial ratios mean for a company is much more difficult to automate. A bank clerk pays out banknotes to a customer – a task that is easy to automate using an ATM and may even be rendered superfluous by the use of cashless payment systems. But the bank clerk also gives personal financial advice to a client who is considering borrowing money to buy a house, and tries to persuade customers of the best approach. These are abstract tasks that are still best done by a human being. The welding work in a car factory is easy to automate, but the assembly of parts is a manual task that turns out to be difficult for robots to manage.

Maarten Goos of the University of Utrecht and his co-authors have carefully calculated the pattern of shifts in employment in sixteen developed European countries since the early 1990s. They compare employment in well-paid jobs such as management roles in large or medium-sized companies and most liberal professions, averagely paid jobs such as office work and skilled factory work, and poorly paid jobs such as unskilled factory work and retail and security positions. In almost all the countries surveyed, the proportion of averagely paid jobs fell between 1993 and 2010 and the proportion of badly paid and well-paid jobs increased.

In Belgium, the proportion of averagely paid jobs fell from 48 percent in 1993 to 36 percent in 2010. This decrease of 12 percentage points puts Belgium in the leading group, together with Ireland and Spain. The proportion of well-paid jobs rose from 34 to 43 percent, and that of low-paid jobs rose from 18 to 21 percent. The figures for the Netherlands are comparable but less extreme.

It is therefore clear that the aim of technological progress – in the form of automation and digitisation – is not only to replace low-skilled and low-paid workers *(skill-biased technological change)*, but above all to replace routine tasks *(routine-biased technological change)*. Research shows that these routine tasks are mainly performed by workers in the middle of the wage distribution with average skills and average wages.

A process of polarisation is therefore playing out in the labour market. Averagely paid jobs are not only the most routine of the three categories, but also the ones that can be most easily transferred to low-wage countries. Quite a few poorly paid manual tasks – the work of a hairdresser or gardener, for example – need to be carried out on the spot and in person and it is difficult for such tasks to be shifted abroad. The figures show that it is mainly routine that is leading to the observed polarisation and contraction of the middle class, with the relocation of jobs abroad being less of a factor.

What workers and horses have in common

The rise of cars and tractors in the early twentieth century destroyed an important economic sector based on the use of horses in agriculture and transport. As a result, the price of horses fell by about 80 percent between 1910 and 1950. This lower price may have slowed the mechanisation of agriculture and transport somewhat, but the number of horses and mules used for work still fell drastically, for example from 21 million in 1918 to 3 million in 1960 in the United States.

Can we draw a parallel between the fate of horses and that of workers? In a way, yes. Estimates indicate that for every additional industrial robot per 1,000 workers, employment in the United States will fall by between 0.18 and 0.34 percentage points and wages will fall by between 0.25 and 0.5 percent. In Europe, the number of industrial robots increased fivefold in 25 years, from about 0.5 robots per 1,000 workers in 1993 to 2.5 in 2018, but it appears that, in spite of this increasing automation, both employment and wages increased in the long term. How can this be? What economic mechanisms are involved in automation?

How automation affects employment

Suppose that machines largely automate the production of tennis balls, thereby cutting the amount of labour. The effect of this automation on employment depends on at least four things: the extent to which the new technology can replace certain tasks (substitutability), the extent to which demand for tennis balls responds to a fall in their price (the price elasticity of demand, in the jargon of economics), the extent to which demand for tennis balls and other goods responds to an increase in income (income elasticities), and the extent to which workers offer more labour when wages rise (the elasticity of labour supply). Let us take a closer look at each of these factors.

If a machine performs routine tasks more cheaply than a typical worker, the price of a tennis ball will fall. Given the same production volume, there will be less demand for those workers who mainly perform routine tasks. We can thus expect their wages to fall. But, at the same time, workers who perform tasks that *complement* the machines will become more productive. Their wages will increase unless so many workers with the right skills immediately want these jobs that they push the wage level down. If that happens, economists say that the supply of labour is elastic. The lower price of tennis balls will cause the demand for tennis balls to increase. If the price elasticity of demand for tennis balls is sufficiently high, the increase in demand will be large enough to increase the total amount spent on tennis balls. This will partially reverse the decline in the employment and wages of workers in the tennis ball industry.

Bodywork construction for the Audi A1 in Audi Brussels

But there is an additional effect: automation leads to higher productivity and lower prices, generally resulting in higher incomes. If these higher incomes sufficiently increase the demand for routine tasks, this may also indirectly cause the demand for those workers who perform mainly routine tasks to increase.

In short, the effect of automation on employment and wages is complex and ambiguous.

Of course, it's possible that we will one day come up with robots that can build anything human workers can, including new robots. If that does happen, the supply of goods and services, including robots, will be limited only by the natural resources needed for production. In 1930, in his essay *Economic opportunities for our grandchildren*, the well-known British economist John Maynard Keynes postulated that future productivity increases would mean that, in a century's time, production would be sufficient to meet everyone's needs. The first problem, then, is that of the distribution of income: who gets the fruits of our labour? Smarter taxes can offer solutions here. The second problem is: how can we use all this free time meaningfully? After all, work plays a crucial role in self-esteem and our sense of belonging to a community.

Paint department at A1 in Audi Brussels

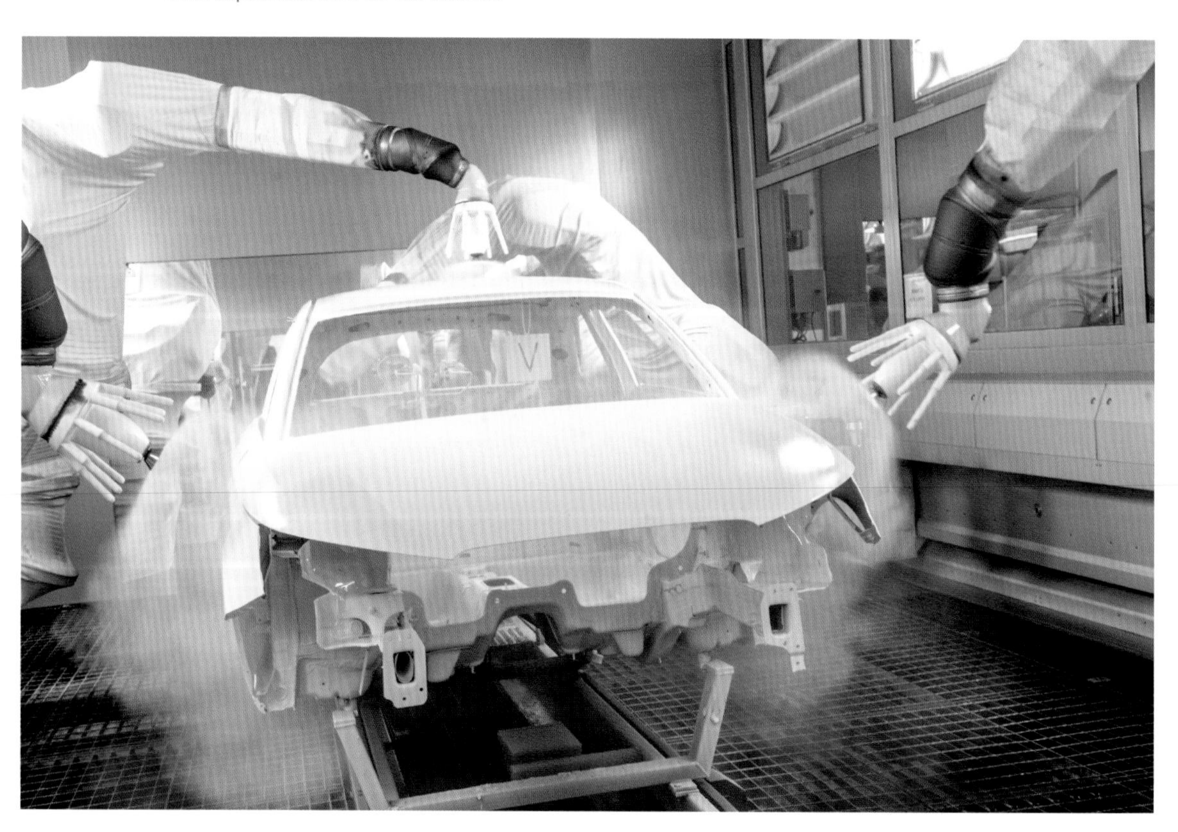

Why it won't come to that

Over the past two hundred years, new technologies have repeatedly led to major productivity gains without a lasting rise in unemployment. Professions such as typesetter or farrier now exist only as niche activities, and we no longer mourn their loss. New jobs have been created by the production of goods and services that we had never heard of just a few years ago. And there are still many unmet needs in the services sector, such as gardening, cleaning, health care and care for the elderly.

Moreover, the impact of information and communication technologies on labour productivity has so far been small. According to American economist Robert Gordon (2016), it is quite possible that the major inventions are now behind us and that the large productivity increases seen between 1870 and 1970 were the exception rather than the rule. That's probably taking it too far. In any case, the actual problem seems to be that productivity growth is lower than we would like it to be.

There are bottlenecks, however

A variety of economic parameters, such as elasticities and consumer preferences, play a key role. Because these parameters are subject to change, it is not possible to make a reasonable prediction about the exact impact of increasing automation or robotization on employment. It is clear, however, that new technologies will continue to cause shifts in the demand for labour. So we need to keep thinking about which skills are scarce and valuable. According to Erik Brynjolfsson and Andrew McAfee, the most important of these are the ability to generate new ideas, recognise patterns in a broad context, and communicate. Education and training are therefore best geared to the development of these skills. But there is nothing to say that these highly-prized skills will remain the same throughout a person's life: lifelong learning is likely to become a necessity.

LELY
Lely Astronaut

The milking robot of Dutch company Lely automatically cleans the udders and attaches the teat cups to the four teats.

©Lely International

WHY DOES ROBOT DENSITY DIFFER SO MUCH ACROSS COUNTRIES?

Of the 40 countries for which the International Federation of Robotics collects data, robot density (the number of robots per 10,000 people employed in manufacturing) varied from 3 in India to 631 in Korea (see chart on next page). The Netherlands, Belgium, Italy, and the United States have quite high robot densities, too (between 170 and 200), but they lag the European robot champions Denmark, Sweden, and Germany (between 200 and 300), and the Asian stars Singapore and Korea (500 or more robots per 10,000 people employed in manufacturing).

Two factors that determine robot density stand out. One is the cost of labour: in Belgium, Denmark, Sweden, and Germany for instance, the labour compensation costs in manufacturing are about US$ 45 per hour, many times more than in India and the Philippines (US$ 2 per hour) where few robots are used.

Another factor is the relative size of the automotive industry: the automotive industry employs more robots than any other sector. This helps explain the high robot density in Korea, Germany and Japan, countries that in 2016 produced between 70 and 80 cars per 1000 people. The automotive industry also explains why the Czech Republic has a robot density that is about double that of Portugal, even if the labour cost in both countries is about the same (US$ 11 per hour): in 2016 the Czech Republic produced a whopping 130 cars per 1,000 people, almost ten times as much as Portugal and more than any other country in the list.

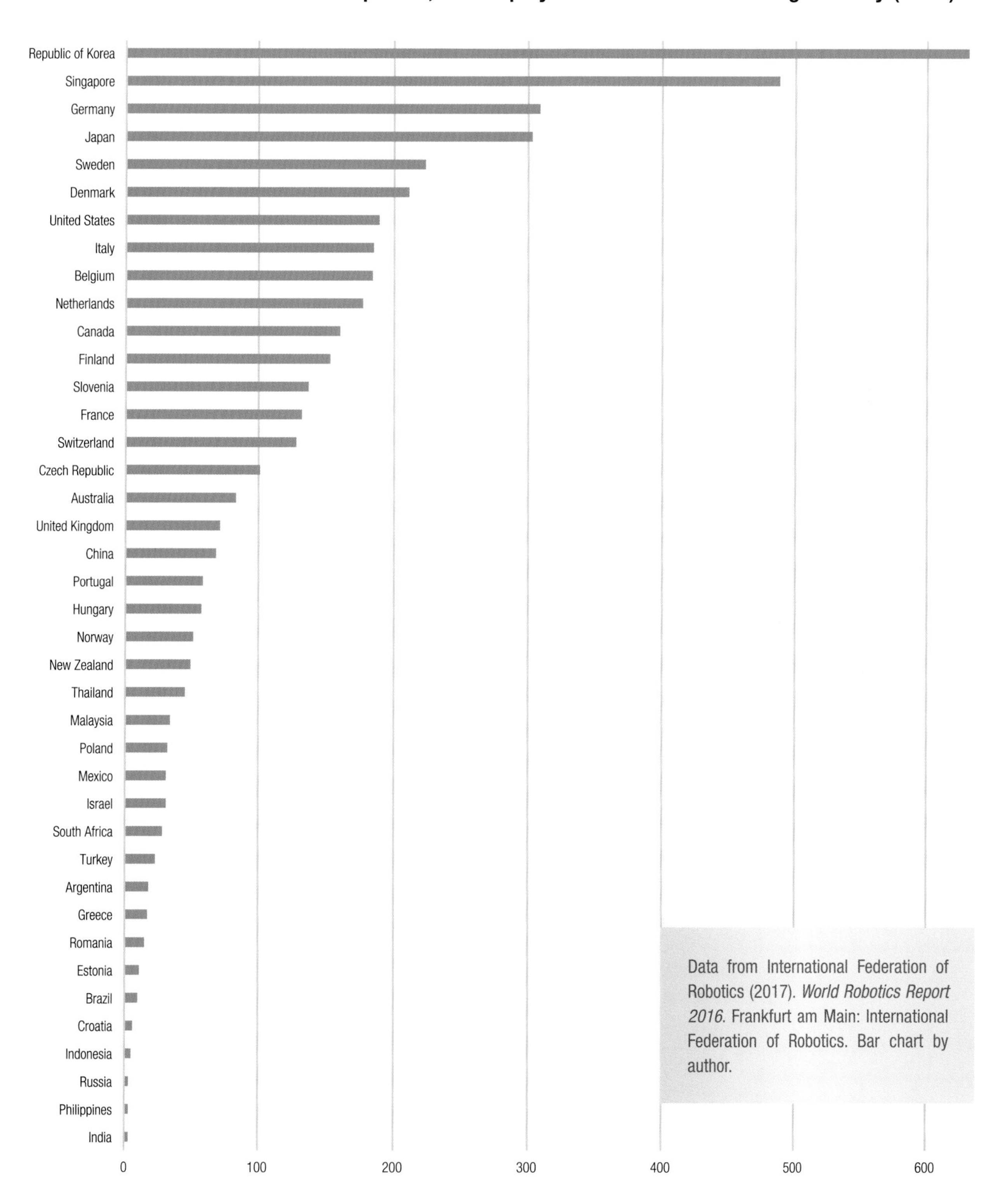

Data from International Federation of Robotics (2017). *World Robotics Report 2016*. Frankfurt am Main: International Federation of Robotics. Bar chart by author.

Based upon the deep-learning software of Robovision from Ghent, the planting robot recognises the stems of the cuttings and plants them in a pot.
©Jonathan Berte

Exoskeleton to reduce physical stresses when working overhead at the BMW plant in Spartanburg.

©BMW

WILL YOU END UP WORKING SIDE-BY-SIDE WITH A ROBOT?

By Prof. Dr. Veerle Hermans, Dr. Ir. Greet Van de Perre and Dr. Christophe Vanroelen

Do you have a heavy job that involves vigourous physical labour? Does your back or neck ever cause you problems after a long day's work? According to the most recent edition of the *European working conditions survey* (2017), more than 25 percent of employees still spend a quarter or more of their working day doing heavy manual handling, more than 60 percent perform repetitive actions and more than 44 percent have to work in tiring or painful postures. Exposure to physical stresses is responsible for a greater proportion of work-related health inequalities among European workers than psychosocial hazards, although the latter currently receive more attention from policy makers. Moreover, exposure to physical risks has remained virtually unchanged over the last ten years: despite all the automation that is around today, there is still a lot of manual work that needs doing which is causing a significant impact on workers' health.

Cognitive workload is increasing

But what about cognitive overload at work? Just over one-fifth of employees report that their job involves highly complex tasks, and two-thirds don't perform any repetitive work at all. Only 15 percent of employees work in jobs where repetitive work predominates.

The labour market has seen a systemic shift from simple, monotonous jobs to complex ones that call for specific competences. If the competences were assigned an index value of 100 in 1995, we are now twenty years further at 125. This means that we are now carrying out increasingly complex work that requires more training.

This decline in repetitive jobs is partly down to a change in production methods. The era of mass production is now at an end, with today's consumers demanding more say in the offerings on the market and the specifics of their chosen products. Customers expect to be able to pick out the components of a new laptop for themselves or select the configuration of their car, such as seat type, bodywork colour and a whole range of options. Companies are striving to meet these new customer desires by transforming a standard item into a large family of products with countless variations. This flexibility in the product range is making production lines more complex as they need to incorporate a range of parts and finishes. This, in turn, is increasing the cognitive load on operators.

We are seeing similar developments in service work. Routine tasks are being automated, and those that require complex decisions, problem-solving and coordination are gaining in importance. This raises the question of how we can provide cognitive support for employees.

Our ageing population is another factor, with the number of 55- to 64-year-olds in employment in the European labour market rising from 16 to 20 percent between 2000 and 2015, meaning that one-fifth of working Europeans are now over 55. Getting older often entails a decrease in

Festo combines the strengths of a human being with those of the pneumatic robot BionicCobot.

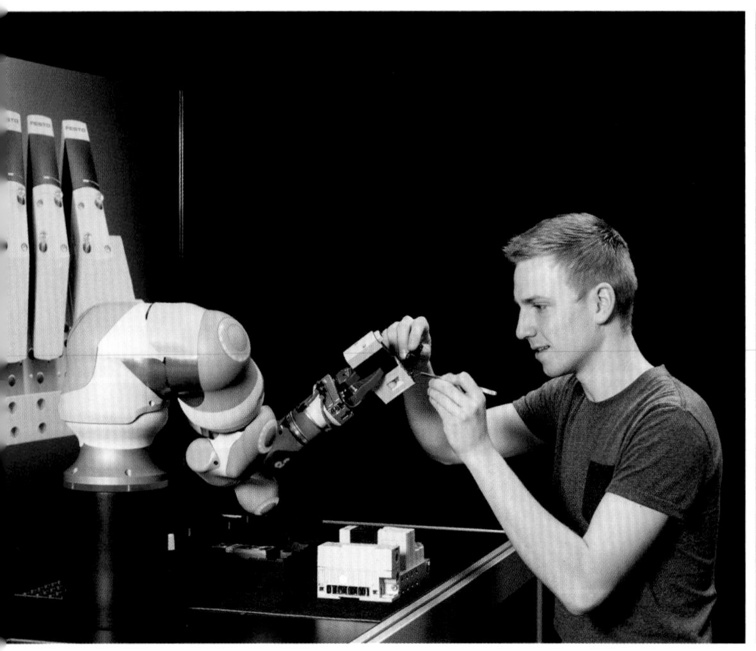

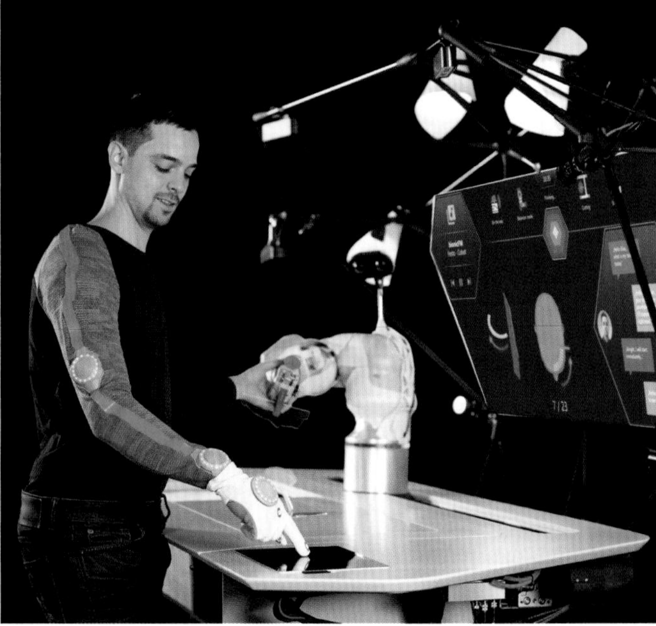

physical capabilities, such as reduced fitness and lower muscle strength and agility. But also cognitive abilities are affected by age, with memory typically deteriorating over time, along with problem-solving capacity and the flexibility to adapt to new tasks. A good ergonomic workplace with sufficient physical and cognitive support is therefore crucial, as well as appropriate preventative measures and thoughts for how to make work manageable for all, regardless of age. So, how can we offer such support?

'Humans are underestimated'

A first step towards relieving physical workload by the use of robots was taken in around 1970, when welding robots were introduced into the automotive sector. Classic industrial robots are mainly used to perform demanding, repetitive or dangerous tasks, with these heavy, cumbersome machines enclosed in cages to prevent hazardous interaction with operators. But not all arduous or dangerous tasks can be taken over by robots and full automation is not a given.

The surgeon at UZ Brussel wears a Microsoft Hololens. Augmented reality provides visual assistance based on MRI/CT scans, helping to ensure the operation goes well. This imec.icon SARA project (Intro-operative Hololens Visual Assistance) is being jointly developed by VUB and Cronos.

©Rafal Naczyk

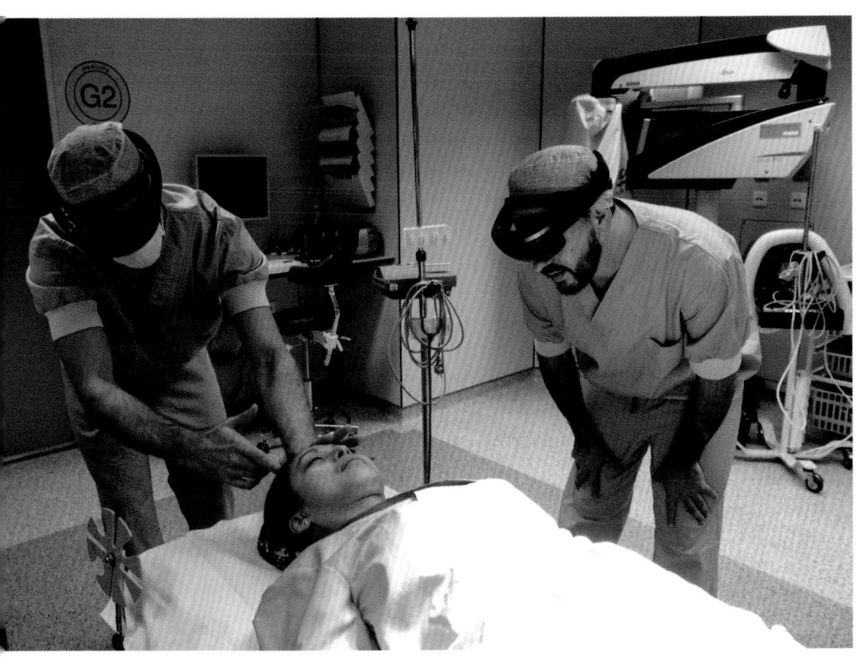

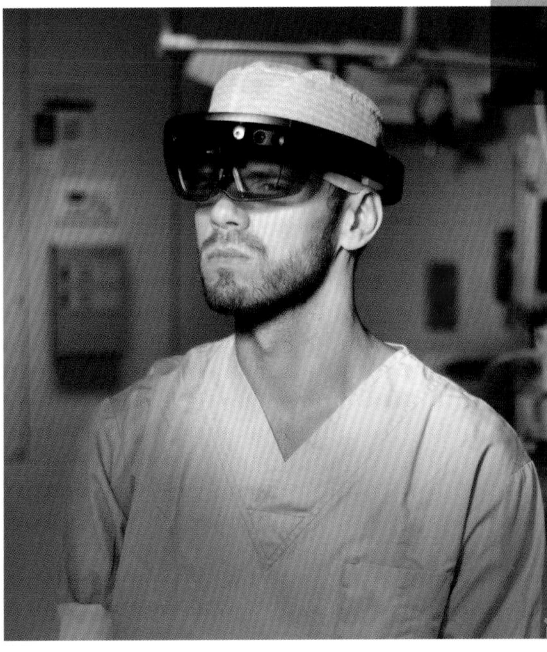

LARGE AND SMALL BOTTLES

A few years back, a cosmetics company had designed everything for the installation of a fully automatic line right down to the last detail: the supply of empty bottles, their automatic filling with body lotion, the screwing on of the pump, the packing of the bottles into boxes and the loading of those boxes onto a pallet. But it turned out that the line needed daily adjustments to meet the customer's requirements, with one wanting 100 ml bottles, the next preferring 250 ml and yet another demanding 500 ml. It took so long to set up and test the system that the company quickly decided to go back to using production operators for the supply and removal of the bottles. Back to square one, and the true masters of repetitive work.

Spexor exoskeleton developed by the Vrije Universiteit Brussel (VUB) to support the spine during heavy physical work. ©VUB-Brubotics

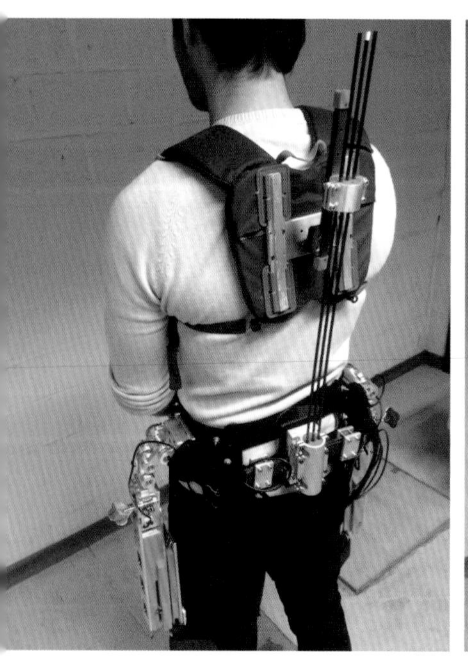

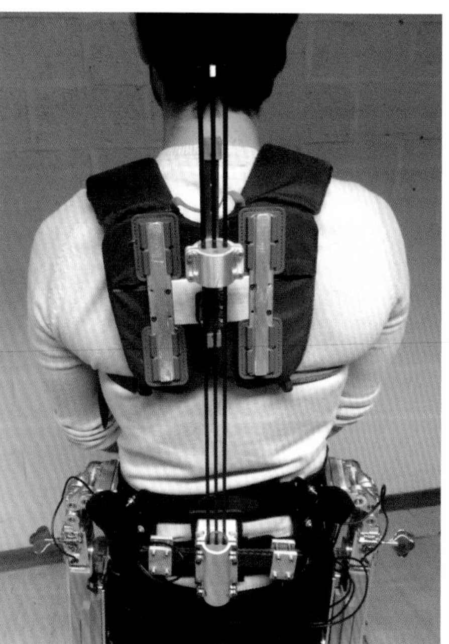

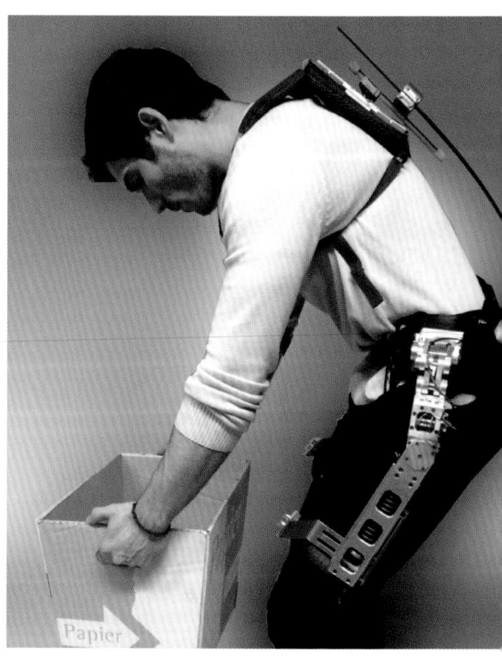

Electric car manufacturer Tesla is no stranger to the problems of automation. To quote CEO, Elon Musk: 'We had this crazy, complex network of conveyors belts. And it was not working, so we got rid of the whole thing.' Later he added that: 'Excessive automation at Tesla was a mistake. To be precise, my mistake. Humans are underestimated.'

And there are yet more examples of places where it is robots, and not workers, that have found themselves out of a job, with Toyota and Foxconn being two that spring to mind. Very often, fantastic systems are invented but when it comes to their implementation in the real world, things don't go as smoothly as was anticipated. Some tasks are by their very nature difficult to automate, with prime examples being anything related to social skills (such as negotiating and dealing with human sensitivities) or cognitive intelligence (such as creativity and flexibility). It's also difficult for automated systems to get to grips with the challenges of perception and manipulation within an unstructured or changing work environment.

Good teamwork is key

Often, therefore, the flexibility, intelligence or problem-solving capacity of a flesh-and-blood human employee is what is needed for the successful performance of a task. So it can be worth considering combining the strengths of a robot with those of a human being. Collaborative robots, or cobots, are robots that are designed to permit safe interaction and collaboration. They don't need to be screened off from human employees as their hardware and software are adapted to allow man and machine to work together. This means that a physical load can be distributed between the cobot and the employee using it. For example, the cobot can hand over or position heavy parts, while the operator performs fiddly operations on them that are difficult to automate. The cobot can also take over part of the cognitive workload, for example by giving the employee instructions or performing quality control.

The wearable robot

Another method of providing physical support for operators is the use of exoskeletons. An exoskeleton is an external, wearable mechanical structure that increases an employee's strength, or limits the impact of a movement or action on their body.

They were originally developed for use in a military context, but were subsequently put to work in the medical sector, where they helped patients undergoing rehabilitation to relearn how to

GIVE US A WINK, WALT

Car manufacturer Audi introduced its first cobot in March 2017 at the Vorst site. The cobot, known as Walt, is the outcome of the Claxon project from imec.icon and is used to glue car parts, which it can do much more precisely and consistently than its human colleagues. A face, an occasional wink and simple facial expressions help make it seem more human, leading to greater acceptance by employees, with operators communicating with it intuitively by the use of hand gestures. The cobot has flexible joints and monitors its surroundings using heat sensors and cameras that can detect depth and colour, allowing it to keep an eye out for its human colleagues. Safe interaction is thus assured.

©Audi Brussels

walk, and in the industrial sector to prevent overload injuries. Today, there are many applications of passive exoskeletons on the market that aim to increase employee comfort.

One is the *Chairless Chair*, an exoskeleton that supports employees in a seated position, allowing them to carry their chair around with them, so to speak. This reduces strain on operators as they no longer have to do their jobs standing up or squatting. Other models support the back during the lifting of loads. For example, the *Laevo* and *BackX S* deploy a spring system to store energy when users bend forward and then release it when they straighten up again.

VUB is also developing a new generation of exoskeletons, with examples being *Spexor* and *Exo4Work*. For shoulder and arm support there are also exoskeletons with armrests such as *Schoulder X* and *Skelex*, which support the natural movement of the arms.

Wearables for cognitive support

Wearables can also be used to reduce cognitive load and increase employee performance by linking useful information to the job at hand and providing it to the employee in real time. One example is *smart glasses*, which not only allow users to see their environment, but also present them with a virtual overlay of digital information.

Antwerp-based Iristick designs smart safety glasses with augmented reality functionalities for industry.

Do these new technologies really help?

So, are these new technologies of any real benefit on the shop floor? In a literature study conducted by Michiel De Looze and his fellow researchers into the effect of exoskeletons, the majority of the studies found decreased muscle activity and lower compressive forces in the lower back. However, these positive effects were mainly observed under laboratory conditions: employees themselves do not always seem to be so positive about the usefulness of such devices in realistic working environments, or about how comfortable they are to wear. There may also be unintended consequences. For example, a back exoskeleton can exert extra pressure on the chest or result in an uncomfortable transfer of the load to other parts of the body.

In addition, it's not enough for the exoskeleton to provide the envisaged physical support: workers also need to accept and actually use the system. So it's important that the employees in question take centre-stage and are actively involved in the design process.

Exoskeletons worn by BMW workers reduce the physical load.
©BMW

And what about cobots? Are they effective in reducing the workload or do they simply cause employees additional tension and stress? This is another field in which attempts are being made to develop human-centred technologies, in which interaction is as natural and intuitive as possible. Cobots can be controlled by speech and gestures and taught to perform new actions by demonstration-based programming. This means that, instead of having to do complicated coding, the employee simply shows the robot what to do.

There are already plenty of companies where this kind of collaboration between robots and operators is bearing fruit, such as Mariasteen in Gits, which produces custom-made goods. Mariasteen is a sheltered workshop where people with mental and/or physical disabilities produce complex products with the assistance of cobots, for example applying sealing mousse and gluing glass covers to light fittings. Working with cobots makes the complex tasks easier, without increasing the associated workload. This allows employees with motor disabilities to perform sophisticated work and deliver products with a higher quality guarantee. Overall, workers enjoy greater job security and the company can carry out a wider range of assignments.

And what about sedentary occupations?

Such success stories are the product of a great deal of research, and this work cannot be carried out without the cooperation of employees throughout the development process. This is the only way to develop sound, usable systems that are effective in reducing the physical and cognitive workload.

But it's also important that the balance is not allowed to tip too far the other way: full robotization would mean eliminating all physical activity from the working day, dooming us to long periods sitting at a desk. The disadvantages of a sedentary life have become increasingly clear in recent years: spending a long time sitting down increases your risk of cardiovascular disease, obesity, diabetes and even premature death – and if that's not enough to convince you, it also increases the padding on your posterior! So a little physical effort is highly recommended.

The robot produces a construction by applying carbon fibres to a membrane from the inside. This is similar to the way a water spider manufactures its underwater web.

©ICD/ITKE Universität Stuttgart

BOT THE BUILDER: CAN HE FIX IT?

By Prof. Dr. Ir. Arch. Lars De Laet and Prof. Dr. Ir. Lincy Pyl

The construction industry is not known for its rapid pace of technological innovation, but the successful construction of high-rise buildings, large sports stadiums, long bridges and complex residential and office projects takes more than craftsmanship and engineering – the latest technological tools are also crucial. Digital evolution has reached the world of construction, with robotization, artificial intelligence and further digitisation all set to play their part in shaping the design and construction process of the future.

Digital (r)evolution

Just a few decades have passed since architects first traded their drawing boards for computers, and this process of digitisation has since then opened up a whole range of new possibilities. Lapping with a razor blade to make small changes to a piece of plaster or having to throw it away and start from scratch if larger changes are needed have become a thing of the past. Digital drawing tools now permit complex projects to be designed entirely based upon parameters and if you want to change the number of columns, the width of a window or the number of floors, a click of the mouse is all it takes to adjust the entire design and send the plans out to everyone involved.

In the past, all plans, sections and three-dimensional visualisations were drawn up independently of each other, but nowadays all the necessary drawings are generated from a single three-dimensional model. This not only avoids inconsistencies between the different plans, but also saves a lot of time. Moreover, a 3D drawing of this kind is ideal for exploring the project on an entirely virtual basis. Project developers can allow their potential customers to take a stroll through the realisation of their dream via virtual reality long before the foundation stone has been laid. You can even use augmented reality to conjure up a visualisation of the final project in context on a tablet or smartphone, for example during a site meeting. And perhaps in the future we will use holograms to model our project in full 3D as we sit around the conference table.

The three-dimensional model only really comes into its own when it incorporates all aspects of the structure, such as its full geometry along with all the materials and techniques used, and the required finish. This is called the digital *Building Information Model* (BIM). Since this model holds all the relevant data in a kind of database it makes it really easy to extract information such as required quantities, or to draw up a bill of materials. This comes in handy if you want to specify the facade surface or determine the number of bricks that need to be ordered for it. The model can also eliminate a lot of problems during the design phase – for example suddenly finding out that the rainwater downpipe has been routed through a beam or column. And if a wall needs to be moved by half a metre it's no problem – all the drawings and measurements can be adjusted and any conflicts detected entirely automatically.

This Building Information Model can be used in the design, construction and usage phases, as well as during the possible repurposing or demolition of the structure. It thus has a *building information management* function. The central maintenance of updated electronic data at all

Team working with 3D print robot.
©Aaron Hargreaves/Foster + Partners (left), ©ICD/ITKE Universität Stuttgart (right)

stages of the construction process not only allows for smoother communication between the architect, stability engineer, structural contractor, technical contractor, etc., but is also useful for optimising schedules and shortening delivery and construction times. Anyone who has ever coordinated the construction of a family home may have experienced the domino effect: one late delivery or one contractor who fails to respect the agreed schedule triggers a series of further delays, often giving rise to a lot of negotiation to get everything back on track. It is expected that, in future, artificial intelligence will be used to manage increasing quantities of data, thereby improving the efficiency of the construction process. Automated and streamlined project management and planning is one example of this approach.

Print your own house

Innovation is making its presence felt not only in the creation of digital building models, but also in the development of new building methods and techniques. *Computer numerical control (CNC)* machines producing goods using *computer aided design* (CAD) and *computer aided manufacturing* (CAM) have also become widespread in the construction industry. Two examples that have been common for years are the automatic drilling of holes in steel columns and the welding of profiles using computer-controlled welding machines instead of manual welding techniques.

One highly mediagenic modern building method is the 3D printing of structures or even entire houses – and you might start seeing these sooner than you think. In 2014, a Chinese construction company managed to print ten houses a day based on cement and construction waste with the homes, which cost less than $5,000, being built up layer-by-layer using a 3D printer. Elsewhere in the world, too, new developments are focusing on the 3D printing of structures, which are increasingly based on sustainable materials. In 2016, for example, a small house was built – or, rather, printed – next to one of Amsterdam's canals using a linseed-oil-based material.

But this technique is especially useful for creating individual construction elements or formwork moulds, as 3D printers allow you to print highly complex geometries quickly and accurately. This design freedom can be useful for constructions where each component is unique or where complex shapes have to be created – for example during the renovation and restoration of ornaments. But complex geometries are sometimes required for modern new-build projects too, and 3D printing offers an efficient solution. Nowadays, for example, it's possible to use this technique to create terracotta bricks, each with a unique shape, or print formwork moulds to size permitting the production of complex construction elements that optimise material use.

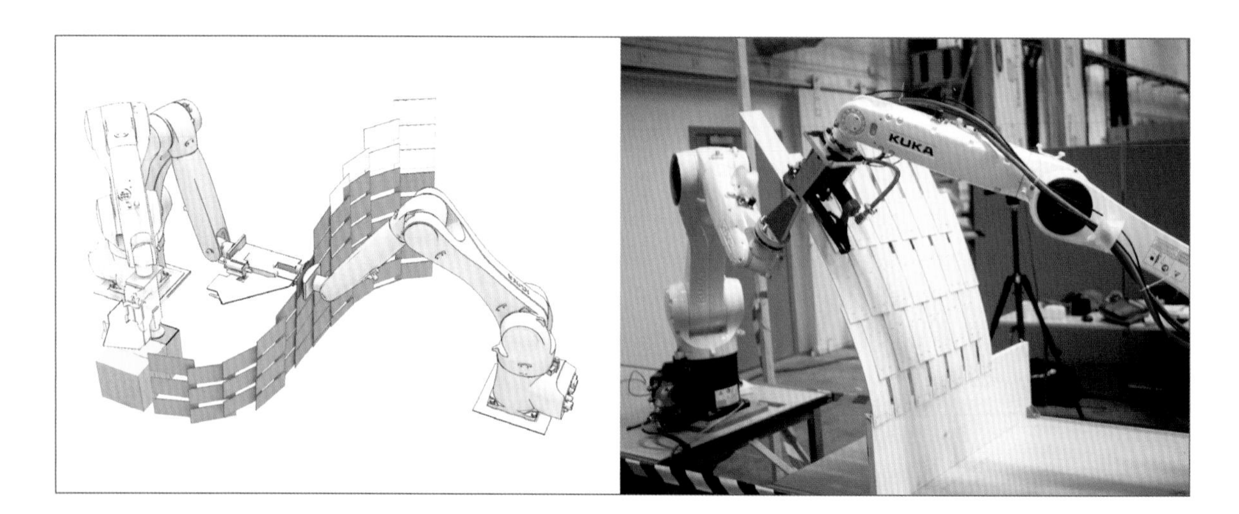

Demonstration of two robots working together to build a three-dimensional structure.
©'Shingled Out', Kyriaki Goti and Erik Martinez-Parachini, Advances in Architectural Geometry 2018, Chalmers University

Recently, researchers in Switzerland developed a concrete floor slab that uses 70 percent less material than a standard full concrete floor slab by positioning the concrete only where it is absolutely necessary from a structural point of view. The floor slab owes its efficiency to its complex and highly specific shape incorporating carefully positioned cut-outs, and can only be manufactured thanks to a 3D printed mould made of sand and resin. Not only does the result look good, it is also much lighter than a conventional slab, meaning that both columns and floor can be thinner, and making it possible to add a few extra storeys to a high-rise building.

3D printing not only makes it possible to produce complex shapes and unique components, it also permits the use of new construction materials. Here too, sustainable alternatives are being explored, such as printing based on ground-up construction waste, sand, clay or natural resins. Soon it will be possible to print using organic materials like fungi and bacteria, thereby converting natural waste streams such as flax, straw and wood into a new and sustainable material. And the new generations of printers are getting better at printing using multiple materials at once, allowing the production of 'smart' components that integrate different properties, techniques and functions (such as conductors, sensors, etc.) into a single component.

Robots and drones at your service

Whereas the use of 3D printing in the construction industry may already feel a bit vintage or passé, the robot and drone are the *new kids on the block* and will undoubtedly have a major impact on the industry in the years to come.

Just as cars are now assembled by an army of robots, in the future building components – and perhaps even entire structures – will increasingly be built by mechanised devices working in close cooperation with human beings. These devices will take the form of mobile mechanical arms onto which various tools such as drill heads, welding machines, nozzles, circular saws, cutters, clamps, and who knows what else, are mounted. A robot arm can also move in different directions and rotate through various angles, allowing these tools to be used in almost any orientation.

For example, one robot can accurately position wooden or steel beams, whilst another bolts or welds them into place. But more repetitive work such as brick masonry is also gradually being taken over by robots, which can place bricks in complex patterns or shapes with extreme precision. And mounting a circular saw or milling cutter on a robot allows extremely complex three-dimensional shapes to be 'sculpted' from solid stone, a method currently being used to manufacture specific parts of Gaudi's Sagrada Familia in Barcelona. It is hoped that this technology will enable the construction to be completed in 2026, a century after Gaudi's death.

Another good example of what we can expect from construction robots in the future is the MX3D pedestrian bridge. This stainless-steel bridge is being printed bit by bit from the two banks of one of Amsterdam's canals. The robots work by printing a section of the overhanging bridge and then driving onto that section and printing the next one. This is repeated until the two halves touch in the middle and the bridge is finished. The bridge's graceful shape is possible thanks to the free-form nature of robotic printing.

We are already familiar with the other *new kid on the block*, the drone, from military applications and as flying cameras and toys (for young and old alike). But these 'flying robots' will increasingly make their presence felt in the construction industry, where they are already being used to take aerial photographs, produce measurements and perform full 3D scans of construction sites. They can't yet lift heavy weights into the air, but once this problem has been solved they will also be able to deliver materials straight to wherever they are needed on site, eliminating the need for a construction crane in some situations.

Robot and drone are coordinated with each other and work as a team to manufacture the structure. The entire production process was planned right down to the last detail in a virtual environment.

©ICD/ITKE Universität Stuttgart

A great deal of experimental work is also currently being done into drones that can automatically pick up building blocks and stack them in the desired place. So, we may soon be able to call upon a fleet of drone-based 'flying builders' to help with the construction process. Research projects have also been successful in the use of drones to stretch carbon fibres between two points, thereby building a roof or bridge.

Robots and drones are currently programmed and steered by humans but, as in other sectors, they are set to become autonomous, communicating and controlling each other using artificial intelligence. Just like the self-driving car, this is no longer confined to the realms of science fiction, with the first self-driving excavators and dumper trucks already in use. Google is also studying the possibility of having its new headquarters built by 'crabots', small self-propelled robots that can operate like a crane to position and move construction elements. In the future, these crabots may allow us to temporarily make rooms bigger or smaller while a building is in use by simply moving floors and walls as easily as we would move a conference table today.

Smart structures

In addition to the robotization of the construction process, our infrastructure and buildings are also set to become smarter. In the future, infrastructure such as bridges and tunnels will increasingly be equipped with a whole range of sensors, allowing them to be remotely monitored on an ongoing basis. Take a moment to think about the chaotic traffic conditions we have endured in recent years due to the temporary closure of bridges and tunnels to assess whether the structure is still safe to use. By performing measurements on a continuous basis, it is possible to determine at any time in the life span of a building, bridge or tunnel whether there have been any changes in the behaviour of the structure that could lead us to suspect damage, allowing the amount of this damage to be measured. Engineers call this *structural health monitoring* or SHM, and it makes it possible to intervene more quickly with targeted and less drastic repairs – which are, of course, much more convenient for the user. Difficult-to-reach structures, such as the inside of bridge girder tubes or wind turbines several kilometres out to sea, can be monitored using sensors within the structure, and the resulting data analysed to reveal when intervention is needed.

At present, such analyses of enormous amounts of data are often still performed by experts, but in time we will be able to replace this human expertise with artificial intelligence. We will teach machines and computers how to interpret the large amounts of data produced, and they will eventually be able to alert us when we need to take action.

Artificial intelligence is also set to become an increasingly tangible presence within our homes. The continuous recording of data so we can learn from it can help us to control or adjust the energy consumption within a home, using smart devices to match lighting and heating to the needs of residents. Just think of the smart thermostats that are already taking note of our lifestyles and using algorithms to adjust the heating and ventilation system to suit. In the future, the indoor climate will be even more accurately governed by an extensive network of sensors in combination with artificial intelligence, allowing heating, ventilation, and the automatic opening of windows or skylights to be controlled based on optimal energy consumption and the needs and lifestyle of occupants. In the short term, you will benefit from lower electricity and gas bills and in the longer term you will be helping to protect the environment.

And comfort is not the only application for home automation. Security-related automation in the home or office has been around for a while now, with CCTV systems and smoke and fire detectors that automatically notify the emergency services if they detect anything untoward being just a few examples. Thanks to our ageing population, and increasing efforts to keep elderly people living independently and out of residential care homes for as long as possible, numerous applications are being introduced that make use of ongoing monitoring and artificial intelligence. For example, there are fall detection systems on the market aimed at elderly people that use sensors to detect abnormal movements that may indicate a fall, notifying the emergency services if necessary. It is expected that increasing numbers of such applications will be developed.

In the future, the 'internet of things' (IoT) will not only be used to link our technical installations to smart meters, it will also connect up our household appliances, lighting fixtures, door locks, cameras, lawnmowers and so on to each other and to our smart devices (smartphone, smartwatch), making them even more interconnected than they are today. The giants of technology and their 'assistants' (the Google Assistants, Alexas, Cortanas and Siris of today) are hard at work on this. They aim to use artificial intelligence to offer you the ultimate in home comforts based on your behaviour pattern and wishes.

Bot the Builder?

The digital revolution is now making its presence felt in the construction industry and this will undoubtedly have a knock-on effect on the buildings and building methods of the future. It's still hard to assess what impact new technologies such as digital building models (BIM), robots, drones and the internet of things will have, but these developments are undeniable.

Construction will likely become more efficient. It's not clear what impact this will have on employment, but one thing is certain: as in other sectors, the type of work people do will evolve over time. For example, more people will need to be trained to work with 3D printers, robots, drones, etc.

It is hoped that the most dangerous and stressful jobs will be the first to be outsourced to technological aids as the construction industry currently has one of the highest levels of accident risk of any sector.

We cannot predict what our buildings and infrastructure will look like in 50 years' time but it is clear that technology is allowing us to use more efficient, sustainable materials and create more complex, material-efficient constructions. And that has to be the ultimate goal: using the new tools of technology to create exciting, high-quality, sustainable architecture.

PRINTING WITH FUNGI

The construction sector is responsible for the consumption of half of all mined materials worldwide and the production of one-third of all waste. If we are to make the construction sector more sustainable so that it is ready to face whatever the future holds, we need to switch to organic construction products and start reusing materials and building components.

An interdisciplinary research team from VUB made up of civil engineers and microbiologists is investigating how materials can be grown organically instead of mined, and how these materials can then be used in construction. The team cultivates organic materials based, for example, on fungi and biological waste that can grow into building components in (3D printed) moulds or that can be 3D printed with the aid of a robot and then grow into efficient construction elements.

IS ROBOTIC SHOPPING SET TO BE THE NEXT BIG THING?

By Prof. Dr. Malaika Brengman, Laurens De Gauquier, Stephanie van de Sanden and Prof. Dr. Kim Willems

For most of us, grocery shopping is a necessary evil that we often dread, with the checkout being a particular source of irritation. But it looks like standing in a long queue with your trolley may soon be a thing of the past. In some shops you can already bypass the checkout, paying for your shopping by simply scanning your purchases with your smartphone. But while self-scanning and self-checkout systems leave customers to do these jobs for themselves, artificial intelligence and cameras may eventually dispense with the checkout process altogether. For example, AmazonGo is a fully cashless shop that has recently been introduced by internet giant Amazon and the rest of the retail sector has been quick to pick up on this trend.

But rapid technological developments and the pursuit of greater efficiency and cost savings mean that the job of checkout operative is under threat, giving rise to panic and resistance from employees. For example, following the announcement of restructuring plans, Carrefour employees sabotaged the self-scan tills because they feared for their jobs. Is there any future at all for human shop assistants, or are they set to become a thing of the past? And what do consumers make of these developments? Will they miss the chat at the checkout and their regular dose of human contact? After all, for many, shopping is not only a chore but also a social event.

Shops without shop assistants

The first unmanned shops are already appearing in towns and cities around the world. You check in using an app on your smartphone and the app automatically takes care of payment, meaning that you never need wait outside a closed shop door again. This shopping concept is currently being trialled in the Belgian town of Woluwe-Saint-Stevens, while the German DIY chain Würth recently opened an unmanned shop where professionals, who are often on the move before dawn, can find the most popular products 24 hours a day. If successful, the concept will be rolled out across Europe.

A shop without shopping assistants is also a potential solution for sparsely populated areas where a manned shop would not be profitable. The first small unmanned Näraffär convenience store opened its doors in the Swedish village of Viken in early 2016. It was the initiative of a resident who found out during a crisis that if you needed to make basic purchases in a hurry you had no option but to go all the way into the city.

Although the retail landscape in Belgium is much more compact, here too local shops are disappearing from the streets, especially in rural areas, and unmanned shops are one possible solution to this problem. In principle, the concept works entirely on trust. Customers scan their shopping with their smartphone and are invoiced for their purchases every month. And, of course, there are security cameras to keep an eye on things.

The humanoid robot Pepper during an experiment by VUB in a confectionary shop, The Belgian Chocolate House, at Brussels Airport.

©VUB-Brubotics

Internet companies want a piece of the action

Internet companies like the Chinese JD.com have already spotted the potential of the concept. Their unmanned JD.id X-Mart stores use RFID technology, artificial intelligence and facial recognition to eliminate the need to scan products. They already have twenty branches in China and recently opened their first store in Indonesia. These are large shops with a floor space of 270 square metres, offering a wide range of products including fashion, cosmetics and accessories. By contrast, the American Shotput is more like an unmanned pickup point for groceries and meals that have been ordered online.

Humanoid robotic shop assistants

We are also seeing an increasing number of trials exploring the extent to which robots can be deployed in the retail sector. An experiment conducted by VUB at Brussels Airport showed that the humanoid robot Pepper in front of a confectionary shop, The Belgian Chocolate House, attracted more attention and had significantly more stopping power than an information terminal. Moreover, the robot appeared to have a positive effect upon shoppers' impression of the store.

If you want to meet a humanoid robot like Pepper in a shop in Belgium, you can visit the European flagship store of electronics giant MediaMarkt in Wilrijk. Don't expect too much – the robot is not yet used in a very functional way and a dance and a map of the shop are about your lot. But it can let you know that there is a traffic jam on the Antwerp ring road and you may as well keep on shopping for a while longer.

American DIY chain Lowe's has been experimenting with the use of its own home-grown service robots in its stores for several years now. The Lowebot greets customers and asks if it can help; it can recognize the desired items visually and guide the customer to the right shelf. This is particularly useful for products you need to do a specific job – as you only buy them occasionally you probably don't know where to find them in the shop. DIY stores are often very big, so the robot can save you a lot of time.

The Carrefour supermarkets in Spain have also been experimenting with humanoid robots on the shop floor for some time. Besides welcoming customers and entertaining children, they also allow customers to fill in a short satisfaction survey. Another useful application is their *wine recommender*, which can suggest the ideal wine to go with your meal.

In the Korean supermarket eMart you can even try out a personal robotised shopping cart. Eli from Robotnik is a tastefully designed self-propelled shopping trolley that guides you through the store using its in-store GPS navigation function. The mobile assistant is equipped with a smartphone holder and an extra-large screen on which you can consult your shopping list and tick off the items as you put them in the trolley. It also offers additional information about the products in the store.

ROBOTS TO THE RESCUE

In Moscow, promobots have been successfully launched in seven hypermarkets of Russian chain Lenta. These bots move through the store autonomously, greeting customers by name using facial recognition software and providing information about discount offers, promotions and new products. One such promobot proved its worth by saving a girl from getting crushed by a falling shelf. When the robot detected the girl climbing on the racking, it rushed over and extended its arm to stop it from toppling over. A neat illustration of the capacity of robots for quick thinking and rapid action.

Shopbot Fabio fired after his first week at work

In an experiment by Heriot-Watt University for the BBC series *Six Robots & Us*, the humanoid shopbot Fabio was used to guide customers around the Scottish supermarket Margiotta in Edinburgh. He was programmed to greet customers and give them directions to hundreds of products in the store, and could also give playful high fives or hugs, and even tell jokes. However, he was dismissed after just a week because he failed to live up to expectations. The robot couldn't understand customers properly and sometimes even scared them off. And his instructions on how to find the shop's products also turned out to be far too generic, with the robot telling people things like 'beer can be found in the alcohol section' or 'cheese can be found in the refrigerators'. Because of his limited mobility, he couldn't move through the shop to guide customers to the products they were looking for. He was banished to a single aisle, where his role was limited to handing out samples, but even there he turned out to be a disappointment. While his human colleagues gave out twelve samples per quarter of an hour, Fabio could only tempt two customers over the same period. It turned out that the customers were deliberately avoiding him.

But when Fabio was sacked, the farewell turned out to be unexpectedly emotional. Devastated, he asked: 'are you angry now?'. Staff were sad to see their new colleague leave. They had become attached to their robotic workmate and some even shed a few tears. What's remarkable is the fact that they didn't see him as a competitor who might threaten their jobs, but as a source of support because he could answer the same boring questions time and time again. Perhaps the robot also inspired empathy because of his human-like appearance.

This experiment shows that, as yet, robots are no substitute for the human contact that consumers are looking for in a shop. As well as carrying out their duties, shop assistants also play an important social role: they know their customers and are happy to have a chat. Robots have not yet taken on that role, although their technology would allow it.

Robots behind the scenes

Experts expect stores to continue to innovate and invest in robotics in the years to come in their pursuit of efficiency, and in an attempt to keep one step ahead of their competitors. Even if robots are not considered appropriate for direct contact with the customer on the shop floor, they are bound to be deployed behind the scenes, where there are lots of jobs in which robots can make themselves useful.

There are plenty of experiments going on to investigate their potential. The Marty robot has an extensive range of duties at various branches of the Giant Food Store in America (part of the Ahold group). He checks whether floors are clean, whether shelves are in need of restocking and whether the displayed prices are correct. But if he finds that something is amiss, all he can do is flag it up so that a human employee can sort it out. This robot does not yet have the judgement and fine motor skills required to stock shelves, meaning that human hands are still indispensable. Another example is at Walmart, where self-propelled robot Emma is put to work cleaning the floor.

In the fashion industry, too, robots can be found behind the scenes. At Sublime Shop, a hip fashion store in Hannut, Belgium, an advanced robotised system is used to fetch shoes from the underground warehouse, freeing up shop assistants to focus on their customers. It can deliver five hundred pairs of shoes per hour, a feat that was unique in Europe when the system was introduced in 2015. And, in the Zara clothing chain, a robotic arm is used at collection points for *click and collect* deliveries.

Processing of online orders

But will we need to visit shops at all in the future? We are seeing a big shift towards online shopping, which may call into question the role of physical stores. Nowadays it's easy for people to buy things over the internet and thousands of robots are already at work in gigantic distribution centres, sorting products and preparing the 100,000 plus online orders every day.

At Amazon, for example, around 3,000 Kiva robots are in use per warehouse where they find the right racks, pick up the products and take them to the order pickers at the unloading stations. This saves these human workers from having to make their way through endless aisles of products, a job that used to be carried out by low-skilled, underpaid workers under poor conditions, giving rise to much criticism. It is now done almost entirely by robots. To ensure that the process can be carried out as quickly as possible, the products are continuously being strategically positioned within the warehouse. This makes the task very complex, as collisions between the robots during the sorting process need to be avoided.

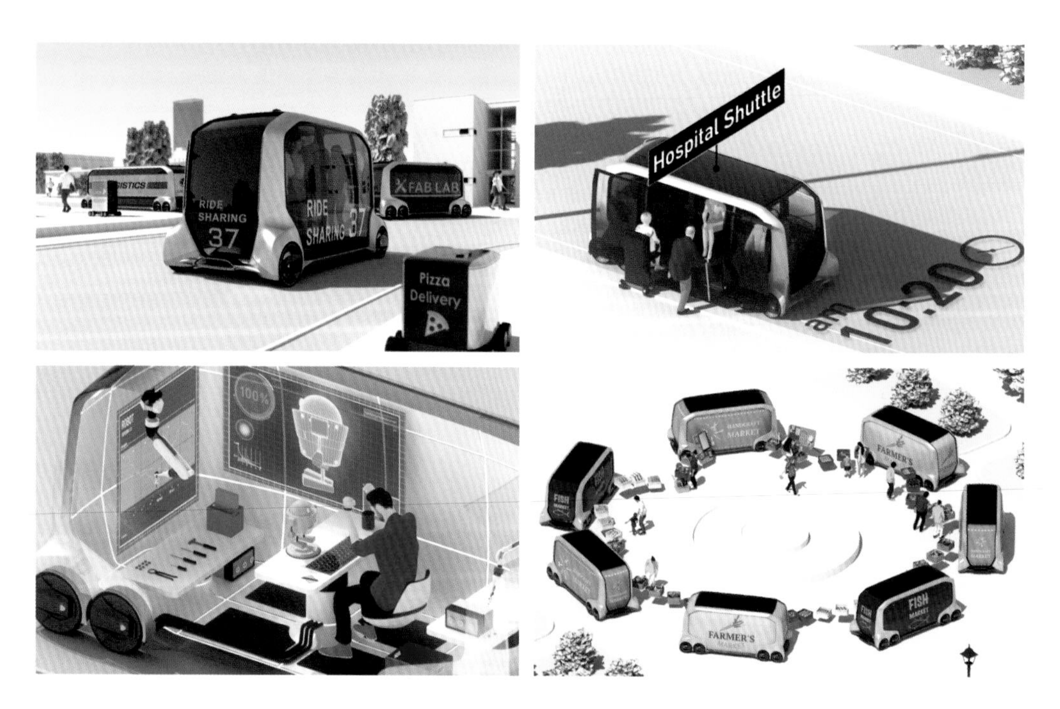

The innovative e-Pallette store concept in which shops in the form of self-driving vehicles come to you. Toyota wants to use the system during the Tokyo Olympic and Paralympic Games.

©Toyota

The final stage, handling the products and making up the shipment, still has to be performed manually by a human being. This is a difficult job for a robot, given the wide variety of shapes, dimensions and materials of the millions of products that have to be processed in the distribution centre. The robot would have to recognise all these products, assess their position and adjust its grip appropriately – another major challenge.

Amazon has even organised a competition for this, the Picking Challenge, in which 28 international teams recently participated at ICRA, the world's largest robotics congress, in Seattle. The task of filling a basket with the desired products seemed simple, but proved to be far from trivial. The recently introduced robots Fetch and Freight work as a team to solve the problem, with one moving rapidly around the warehouse and the other, larger, robot using its arm to grab objects from the racks. However, this research project is still in its infancy, and the capabilities of the robot arm remain very limited.

Home deliveries

Autonomous robotic vehicles are also increasingly being used for the home delivery of products purchased online. It's not only Amazon that is experimenting with the use of self-propelled robots and drones to get goods to their destination as quickly and efficiently as possible – Domino's Pizza is looking into this too. And in Scottsdale, USA, the supermarket chain Fry's Food Stores (from Kroger) has launched a pilot project to deliver groceries using self-driving vans.

If you prefer to do your own shopping in a physical store, there are also mobile shops that can come to you without any human intervention. At any rate, that's the concept behind Wheelys MobyMart, which was developed in collaboration with a Chinese university. A prototype of the mobile unmanned robotic supermarket – which makes its way to you autonomously when you call it up via an app – is being tested on the streets of Shanghai. It's there for you 24 hours a day and offers a range of products for immediate consumption, such as milk or lunch, as well as over-the-counter medicines. You use your smartphone to scan and pay for your shopping.

Japanese car maker Toyota recently introduced e-Palette, a self-driving shopping platform that supports e-commerce. It's a kind of minibus-cum-shop that can find its own way to you to bring you your shopping, and even lets you try on the shoes or clothing you have bought online. Toyota has achieved this by collaborating with the big guns such as Amazon, Uber, Pizza Hut and the Chinese company Didi and is hoping to demonstrate its fleet of multifunctional mobile shops during the 2020 Olympic Games in Japan.

So, what does the future hold?

Experts believe that the market for robots in the retail sector is set to continue growing steadily over the coming years, with a market analysis by leading global consultancy firm Roland Berger predicting an increase from 19 billion dollars in 2015 to 52 billion dollars in 2025. This development will have a major impact on the way we do our shopping in the future.

But we shouldn't forget that consumers will have an important say in this matter. Because, as demonstrated by the sad tale of Fabio the mechanical shop assistant, if consumers don't take to this new technology they will simply show the robots the door. Ultimately, it's a case of he who pays the piper calls the tune: the customer is still king. If robots can offer consumers something extra they may welcome them as part of their shopping experience but, if not, there will still be jobs for human shop assistants at the checkout.

The robot Fetch consists of a robotic arm that gets products down from a rack and places them in the basket of its colleague Freight.

©Fetch Robotics, Inc.

HEALTH

Kevin Andrew Evison of ARM Team Imperial during Cybathlon, a competition for bionic athletes.

©ETH Zurich / Alessandro Della Bella

WHAT IF ATHLETES TURN INTO ROBOTS?

By Prof. Dr. Romain Meeusen, Sander De Bock, Jo Ghillebert, Prof. Dr. Kevin De Pauw

OLYMPIC GAMES FOR CYBORGS

The Cybathlon is a high-tech – and thus, according to its fans, clearly hipper – version of the Paralympic Games in which a helping hand from technology is not against the rules. Quite the contrary, in fact. Athletes compete with each other using motorised prostheses, or are assisted by exoskeletons. The first Cybathlon was held in Zurich in 2016, attracting 66 participants from 25 countries. The disciplines included a brain-computer interface race and a functional electrical stimulation bike race. Belgium fielded two teams supported by scientists from the Brubotics consortium of the Vrije Universiteit Brussel (VUB).

Another such competition is the 'RoboCup', the robotic version of the football world cup. In other words, cyborg games are now a matter of fact rather than fiction.

Lessons and strategies from history

The motto of the Olympic Games is 'citius, altius, fortius', or 'faster, higher, stronger'. Athletes are constantly trying to push their limits by using tricks and gadgets to improve performance. The difference between a bronze medal and last place in the final of the 1500 metres during the

Olympic Games in Beijing was very small. This also illustrates that the other philosophy linked to the Olympic Games – 'participation is more important than winning' – is an empty one.

In ancient times, Olympic athletes were considered heroes and even then were constantly striving to improve their performance. One example is Milos van Croton, who as a young boy trained by carrying a sheep on his shoulders. Later on, he carried a calf and, at the peak of his ability, an adult bull. These ancient heroes also experimented with protein-rich food and 'special' mushrooms.

At the Athens Olympics in 1896 – the first 'modern' Olympic Games – the marathon was won by Louis Spyridon in a time of 2h58'50". The current world record stands at 2h01'49" (Berlin 2018) and last year an attempt was made to break the magical two-hour limit by a number of African runners supported by a group of scientists. Every technical trick in the book was brought to bear – from ultra-light footwear to a 'pacing car' to individual nutritional strategies – but, sadly, the Kenyan Eliud Kipchoge still completed the marathon with a time of 2h00'25". That's still a new 'unofficial' world record though.

So it is possible to push the boundaries. History shows how better training methods have gradually been developed, giving some athletes a competitive advantage. Paavo Nurmi was a Finnish athlete who used the *fartlek* training method (a kind of interval training) for the first time. He won his first gold medal at the Antwerp Olympic Games in 1920. Over a period of twelve years (three Olympics) he won nine gold and two silver medals in the distance running events, before being excluded from the 1932 Olympics (Los Angeles) for 'professionalism'.

From training to technology

Training methods are important, but technological advancements might also enhance exercise performance. Exercise physiology has evolved into a discipline of 'to measure is to know'. Nowadays we are all familiar with heart rate monitors with a built-in GPS function. This is a type of watch that provides information about average speed, distance, intensity of effort, altitude, and so on. This type of equipment is also used in team sports such as football to record training activities, mainly with the aim of generating a detailed report on the stress on players and thereby avoiding overload. In baseball, tiny biomechanical sensors built into a bracelet are used to track the acceleration and accuracy of the pitcher's throwing arm.

A 'smart suit' is sportswear that is not only beautiful, but is also constantly recording data. Examples are the innovative skating and ski suits that aim to optimise biomechanics and thereby minimise air resistance. Samsung has developed a smart suit for short-track skaters that improves the skater's position during training. Short trackers need to position their body as low as they can when skating so that they transfer as much power as possible from their legs to the ice. The smart suit is custom made and incorporates five sensors. During training, the sensors calculate the distance from the skater's hip to the ice and the coach and movement scientist use an app to see whether the skaters are well-positioned – all in real time. The coach can use the app to transmit a vibration that the skater feels on his wrist, telling him he needs to change his body position. This method was used by Dutch short-track skaters in their preparation for the recent Winter Olympics. The two skaters who used this type of smart suit won gold in the 1,000 metres and silver in the 1,500 metres, and their relay team won a bronze medal. This year they were named short trackers of the year.

Physiological measurement systems are being increasingly incorporated into T-shirts. In the run-up to the Tokyo 2020 Olympics, an entire industry is developing the so-called smart textiles that can carry out all kinds of physiological and biomechanical measurements. This will be important given the high temperatures and humidity during the summer months in Japan. These smart textiles not only record information, they also contain specific 'channels', for example to monitor sweating. This allows the development of an individualised schedule to avoid dehydration. The composition of sports drinks is also prepared individually on the basis of previously gathered information.

These smart textiles will soon be next generation heart rate monitors. Tissue blood flow, breathing frequency, fluid loss and the like are fully integrated, and they can also measure 'metabolism'. This data monitoring method allows to efficiently and quickly replace the 8,000 to 9,000 kcal of energy consumend during a mountain stage in the Tour de France. (To put it in perspective: 9,000 kcal corresponds to about 76 bananas or 24 Big Macs, and you won't be able to consume this amount of energy between two days of competition.)

This technological progress will beneficially influence exercise performance, since it will become easier to predict when athletes will go into 'error mode', and can help to avoid overload during training or in competition. The prediction of fatigue and possible overtraining is also set to become increasingly reliant on technological measurements, for example by the instant analysis of saliva samples for signs of overload of overtraining.

Oscar Pistorius, the Blade Runner, during the Olympic Games in London, 2012.
©Will Clayton

London 2012
2790

Are there limits to the benefits of technology?

A nice illustration of technological progress can be found in skating, where the introduction of the 'clap skate' led to a huge leap forward. Gerrit van Ingen Schenau, an expert in biomechanics from the University of Amsterdam, developed a skate that allows to keep the blade in contact to the ice for a longer time, making it easier to push against the ice. This allows skaters to achieve greater mechanical efficiency in their movements. This clap skate is now widely used. New members of the team of Professor Ingen Schenau further optimised the design of the skate with a small mechanical part, which triggered the release of the blade mechanism after several contacts with the ice after the start. However, the device was banned because of the presence of a battery.

Besides the presence of batteries, there is also such a thing as 'mechanical' doping. In 2016, Belgian cyclo-cross rider Femke Van den Driessche was caught with a motor in her bicycle at the World Championship and suspended for six years. In total, only a handful of athletes have been caught, but it is assumed that some of the top athletes from the beginning of this decade owed their 'superhuman' performance to mechanical doping. Therefore, nowadays both athletes and their equipment undergo doping tests.

The Olympic dream of Blade Runner

Oscar Pistorius was born without fibulas (calf bones) and doctors advised his parents to amputate his legs below the knees before he could walk. So he learned to walk using prostheses, growing up as a sporty young man. When he decided to focus on athletics, he was soon winning all the sprint races, even against his non-disabled peers. His big dream was to participate not only in the Paralympic Games, but also in the Olympics itself. It soon became clear that, partly as a result of his special carbon-fibre prostheses, or 'blades', he had developed a special technique that allowed him to sprint more efficiently. After Pistorius achieved extraordinary performances at the 2004 Paralympic Games, he aimed at his big dream, i.e. to compete in the Olympic Games. Top scientists like Peter Bruggeman from the Sports University of Cologne investigated the advantages and disadvantages of Pistorius' blades compared to non-disabled sprinters. They found that Pistorius used 25 percent less energy at the same (top) speed and he was therefore excluded from the next Olympic Games. However, further research showed that, particularly in the first part of a race, he was also at a disadvantage because he can't accelerate to his maximum speed as fast as his opponents. As a result, he won his legal battle and entered the London Olympics. He was part of the South African team that reached the final of the 4x400 meter relay and he himself also ran the semi-final of the individual 400 meters.

Mechanical doping

The story of Pistorius shows that ethical issues may also arise when talking about 'normal' and 'cyborg'. If these blades could be advantageous during sprint races, can athletes have their lower legs amputated and replaced by blades, or would that make them guilty of mechanical doping?

In some sports, competitors are allowed to wear a brace to support an injured joint. Exoskeletons are being used more often in industrial settings. Since exoskeletons are assistive devices, employees carrying heavy loads may have to work for a longer time before retiring. Exoskeletons have also been developed that allow soldiers to carry heavy loads during field operations.

What if these exoskeletons, braces and prostheses provide an advantage in sport? The athlete of the future could use a motorised brace to run faster, jump higher and use energy more efficiently during long-distance races. And what about weightlifting? We are used to seeing the heavy weight lifters tightening their leather lumbar belt before attempting a huge load. Perhaps lightweight athletes could lift the same weights as the hulks if they were assisted by a sophisticated exoskeleton.

Will there be new competitions in which these types of technological advancements are allowed? Even now, athletes in the Paralympic Games try to get into a 'higher' category with the minimal allowed disability. The number of categories based on athletes' disabilities is already enormous. What if we also had to create extra categories for able-bodied athletes? This would require a major effort of sports federations and bodies such as the International Olympic Committee. Rules of the novel competition need to be described. To what extent will these technological advancements be allowed?

Clearly defined criteria already exist for doping. A substance or method can be placed on the doping list if it meets at least two of the following three criteria: (potentially) performance-enhancing, (potentially) harmful to health and contrary to the 'spirit of sport'. The latter criterion raises a large number of ethical issues.

If assistive devices are allowed, we might think about mixing all competitions with mechanical 'gender correction' in those competitions where the physiological difference between men and women is too big. A lot of things need to be considered here.

Michel De Groote during Cybathlon using the VUB robotic prosthesis Cyberlegs.
©ETH Zürich / Nicola Pitaro

Cyborg games: fiction no longer

Oscar Pistorius can't produce his top acceleration over the first few meters of a sprint because his blades don't integrate mechanical parts, but research into new prostheses is moving towards powered devices. Brubotics, a consortium from the Vrije Universiteit Brussel (VUB) that investigates human-robot interaction, is one of the pioneers in this field of research. Several 'assistive' prostheses have already been developed that make it easier for individuals with an amputation to perform everyday movements, such as going up and down stairs or walking on a inclined or declined surface. This is a typical example of the integration of humans and robots, or the 'homo roboticus'.

Cybathlon is an example of a competition in which people with disabilities are assisted by exoskeletons, powered prostheses and other assistive devices. In 2013, Robert Riener, a professor at Zurich University of Technology, launched an initiative in which people with disabilities compete against each other using the latest technological devices. The first Cybathlon competition was held in 2016 and attracted 66 participants from 25 countries. Participants compete against each other in disciplines such as the Brain-Computer Interface Race, in which computer games are played via brain stimulation, and the Functional Electrical Stimulation Bike Race, a race in which riders whose lower limbs are paralysed compete by the artificial stimulation of the nerves that control the muscles of the legs. There are also specific competitions for powered exoskeletons and prostheses for the upper and lower limbs. These usually consist of everyday tasks and movements such as climbing stairs and opening a door. Brubotics VUB competed in the powered leg prostheses race.

There are also genuine competitions between robots, such as 'RoboCup', the robotic version of football's World Cup. Football was chosen because it involves so many aspects of human movement, making it possible to determine how human the humanoid robots and their movements really are. And a 'ski robot challenge' was held during the 2018 Winter Olympics in Pyeongchang, South Korea. It was won by Taekwon V from MiniRobot Corp, the smallest robot of the competition at a height of 75 centimetres. In total, the robot used 21 motors to coordinate its movements, as well as cameras and sensors to detect the correct ski run. However, it remains clear that there is still some work to be done before these robots approach human performance.

It should also be mentioned that robots are already capable to conduct complex movements: a Boston Dynamics robot can execute a back flip and football robots can make 'tactical' decisions based on the movements of their opponents. But they still have significant limitations and cannot yet get anywhere near to mimicking the activity of 800 muscles of the human body working together in harmony. They are definitely a long way off achieving the highly refined motor skills required in some sporting disciplines.

Relevance for society

Elite sports and top-notch technology therefore have a lot in common. Their shared goal is to live up to the Olympic ideal of *citius, altius, fortius*. Will this exceed the limits of the human being? In the future, will there be a homo roboticus competition in every sport? Just as, in Formula 1, technological progress improved the performance of the racing car, it is likely that scientific developments will help to improve general living conditions. Technological developments that are tested in elite sports can lead to spinoffs and thus might become available to the broad public. For example, nanotechnology is increasingly being used in targeted drugs to attack cancer cells, and the technology of implants such as hip and knee replacements is becoming increasingly sophisticated.

Only time will tell whether the Olympic Games will become a competition in which homo roboticus gladiators compete against humanoid robots.

Mark Daniel of team IHMC is paralysed due to a spinal cord injury, but thanks to the exoskeleton he is able to overcome several obstacles during the Cybathlon race.

©ETH Zürich / Nicola Pitaro

The Lokomat (Hocoma) can be used to help patients relearn how to walk during rehabilitation.

©Johan Swinnen

DO PHYSICAL THERAPISTS AND ROBOTS WALK HAND IN HAND?

By Prof. Dr. Eric Kerckhofs, Prof. Dr. Eva Swinnen, Nina Lefeber, Emma De Keersmaecker and Dr. Erika Joos

Rita B., a 63-year-old retired accountant, was rushed to a regional hospital after a stroke. A blood clot had blocked a blood vessel in her brain, killing a number of brain cells and leaving Rita with partial paralysis of her left arm and leg. Two weeks later, she was transferred to a rehabilitation centre. When the rehabilitation specialist examined Rita, the loss of strength in her left leg had left her unable to support herself on that leg, and therefore unable to walk. She could move her left arm a little, but couldn't use her left hand, and was referred for physio-therapy and occupational therapy. With a knee brace stabilising her knee so that she could put her weight on her paralysed leg, it took two physiotherapists to help Rita relearn how to walk, one to help her balance and one to move her paralysed leg forward one step at a time. This gait training, as it is called, was thus an exhausting activity, both for Rita and for the two physio-therapists assisting her. However, the rehabilitation centre has recently acquired a modern Lokomat walking robot, which will make this process easier for both patient and physiotherapist.

The EksoGT (Ekso Bionics) helps patients to walk in the rehabilitation centre.

©Johan Swinnen

What do these walking robots actually do?

To improve safety and stability, Rita wears a kind of corset that uses a pulley system to support part of her body weight, reducing the load on her legs when walking. The robot also incorporates an exoskeleton with straps attached to her upper and lower leg, allowing the robot to move these parts of her body using electric motors, thereby creating a normal-looking gait pattern.

The advantage of this robotic system is that it allows the patient to walk with a normal step pattern for half an hour. Once the patient has been installed in the robotic system, the physiotherapist only plays a supervisory role, meaning that both the patient and physiotherapist reap the benefits.

One ingenious feature of such systems is the fact that the degree of assistance provided by the robot can be adjusted to the patient's abilities. In addition, the walking robot is connected to a treadmill, which allows patients to start gait training at an early stage of their condition – in Rita's case, soon after her stroke. As soon as the patient has sufficient trunk stability, a mobile robotic system can be deployed to help them walk around.

Which patients are suitable for robotic rehabilitation?

Optimal rehabilitation requires four things: (1) lots of practice, (2) sufficient intensity, (3) task-oriented exercises and (4) active participation of the patient. These four aspects are important for both musculoskeletal problems and disorders of the brain or spinal cord. Modern robotic devices offer a wider range of options for fulfilling these requirements than is the case for traditional treatments. Robotic therapy is used for children, adults and older people. Research has been carried out involving children with central movement disorders (for example, cerebral palsy), stroke patients and those with Parkinson's disease or paralysis due to an injury to the spinal cord. Based upon a principle known as neuroplasticity, repetitively training the patient to perform the most correct step pattern encourages changes in the brain or spinal cord. Robotic therapy can also be used for disorders of the musculoskeletal system, for example after amputation of the lower leg.

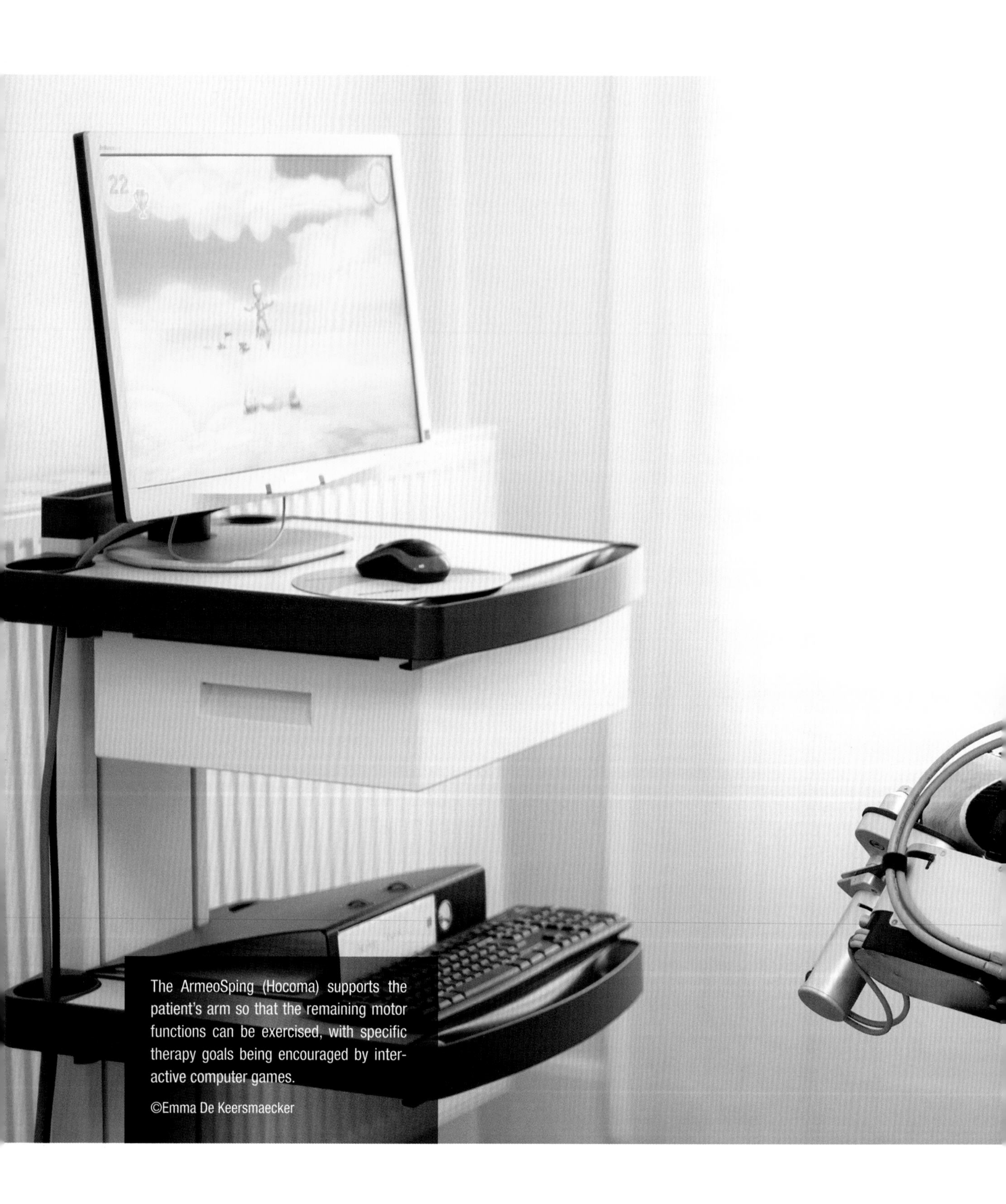

The ArmeoSping (Hocoma) supports the patient's arm so that the remaining motor functions can be exercised, with specific therapy goals being encouraged by interactive computer games.

©Emma De Keersmaecker

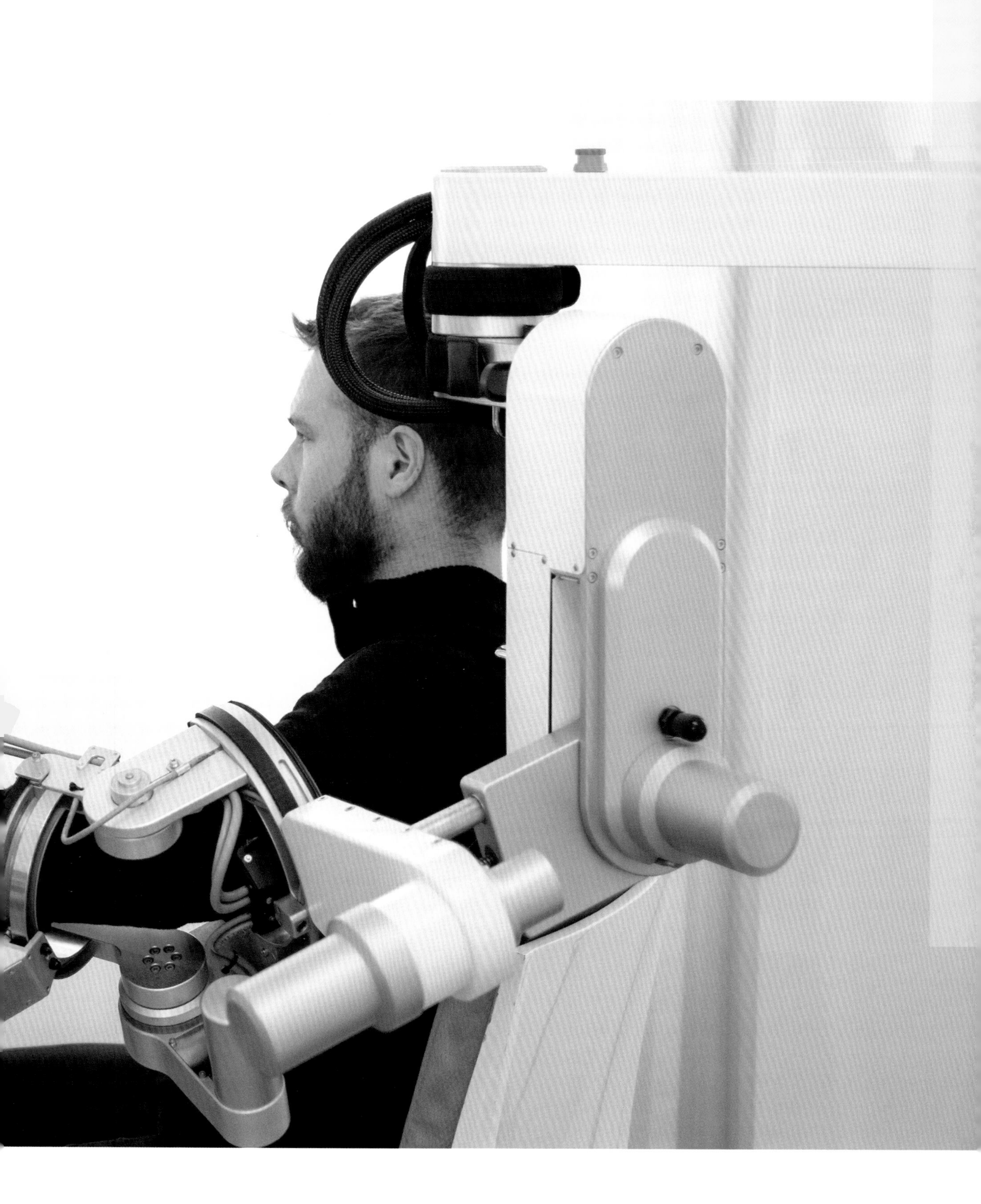

How do patients feel about being treated by a robot?

Patients respond differently to the use of robots in rehabilitation. Recent research in people recovering from stroke in a number of Flemish rehabilitation centres has shown that most patients using a walking robot are highly motivated: in general, they have high expectations of the added value of this high-tech equipment, and these high expectations have a positive effect on the treatment for psychological reasons. However, there are also patients who are a little afraid of these somewhat daunting devices and are less keen on using them. Proper explanation and guidance can help, but also watching other patients can be reassuring.

Can robots replace physiotherapists?

Robots can be a useful part of a treatment program and can help physiotherapists do their job. In the systems that are currently available it takes some time to get the patient installed in the robot and adjust the settings, but once the robotic system has been properly set up, the time it takes to turn it on and off for subsequent sessions is acceptable.

During training, only one physiotherapist is needed for supervision or guidance compared to the two or sometimes even three during conventional gait training on a treadmill. A robotic system thus simplifies the work of the physiotherapist, allowing the patient to exercise under controlled conditions for half an hour or more.

Even though artificial intelligence will be able to offer more substantive assistance in the future, the physiotherapist will continue to be responsible for organising the treatment programme and will always remain involved.

One major limitation: the price tag

In the near future, robotic systems are likely to allow patients more freedom of movement and improve their everyday functioning. This will increase their autonomy – which, of course, is the ultimate goal of rehabilitation. However, the high cost of these systems is currently a major stumbling block and the reason why only a limited number of centres in Flanders have such equipment, with patients sometimes being required to make a substantial financial contribution. This is a significant brake on the use of rehabilitation robots. Luckily, the price of robotic systems is expected to fall in the future and it is hoped that health insurance will intervene.

What does the future hold?

The use of technology in rehabilitation, including robotics, is set to become increasingly common in years to come, with one important source of knowledge being the world of gaming. So-called 'serious games' will be used to increase patients' motivation and allow them to practice in a goal-oriented manner, and also virtual reality will have a role to play. Our knowledge of control algorithms is growing, actuator motors are becoming smaller, and exoskeletons are getting even lighter. All this will undoubtedly lead to better and increasingly functional robotic systems.

One fascinating new development is the discovery that electrical signals from the brain can be detected on the skull and used to control robotic systems via computer programs, with the patient simply thinking of a movement and the exoskeleton carrying it out. The rapid development of artificial intelligence will lead to ingenious, learning robotic systems that can be used to help improve people's lives. Robotics in the service of man, working hand in hand with the physiotherapist!

MENS SANA IN CORPORE ROBOTICO?

Everyone knows the old Latin saying 'mens sana in corpore sano' or 'a healthy mind in a healthy body'. But what few people know is that this saying is actually part of a longer sentence, written by the Roman poet Juvenalis, 'Orandum est ut sit mens sana in corpore sano', meaning 'You should pray for a healthy mind in a healthy body.' Instead of praying, modern technology has produced all kinds of tools to support the human body where such assistance is needed, such as artificial joints, protheses, robot-controlled exoskeletons, etc. In her book *Our Strange Body*, Dutch philosopher Jenny Slatman asks how much of a person's body we can alter or replace before that person loses their sense of identity. This line of thought is relevant to the psychological aspects of the use of robots in rehabilitation, and it is a perspective that we should not lose sight of: a healthy mind in a body with robotic support!

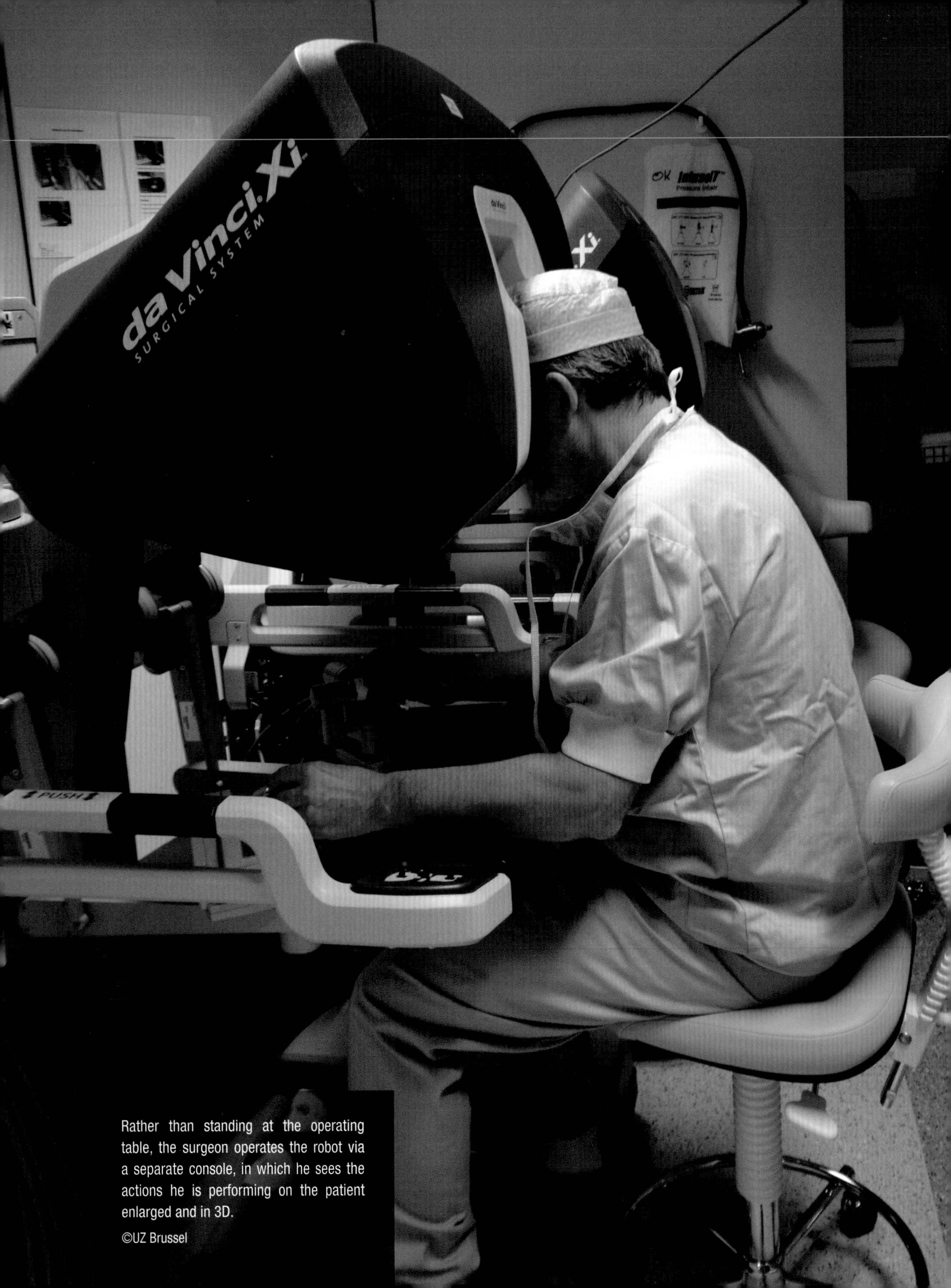

Rather than standing at the operating table, the surgeon operates the robot via a separate console, in which he sees the actions he is performing on the patient enlarged and in 3D.

©UZ Brussel

DO ROBOTS MAKE GOOD SURGEONS?

By Prof. Dr. Philippe De Sutter, Prof. Dr. Marc Noppen,
Shirley Elprama and Prof. Dr. An Jacobs

DEVELOPED FOR WAR VICTIMS

Robotic surgery was originally developed for military use to allow operations to be performed remotely on victims of war. This concept was not taken any further due to the slowness of data connections at the time, but it was subsequently picked up on and developed for medical and surgical applications by the American company Intuitive Surgical. The first robotic surgery system approved by the FDA in the US was launched onto the market in 1999 under the name da Vinci. Such systems are now in their fourth generation, with a reported 4,528 of the various types in operation worldwide. Five million robotic procedures have already been performed using this approach – 875,000 of them in 2017 alone – and this figure is increasing by 15 percent per year. Despite its small size, even Belgium currently has 37 systems in place.

'Robotic surgery' is actually a misnomer as the robot never acts autonomously: strictly speaking, it is a 'master-slave-telemanipulator system' rather than a real robot and it would be more accurate to describe the procedure as 'robot-assisted endoscopic surgery'.

The system consists of a column and four motorised arms suspended above the patient that can move in all directions. These arms are coupled to 8 mm long trocars – thin, hollow tubes that are inserted into the abdominal cavity or chest of the anaesthetised patient through four small incisions. The surgeon, who is seated at a console next to the patient, looks at a three-dimensional high-definition screen that displays images from within the patient's body and uses these trocars to insert an endoscopic camera, as well as instruments for cutting, burning or stitching. The surgeon controls the robot's arms – and therefore the instruments – via a complex system of joysticks, buttons and foot pedals, a skill that requires extensive practice and experience. Inside the patient, the tiny instruments can move in all directions with great precision and free from any vibration.

The da Vinci robot in the operating theatre of UZ Brussel has four arms that are suspended above the patient and can move in all directions. Various instruments are attached to the robotic arms, which the surgeon controls using a complex system of joysticks, buttons and foot pedals.

©UZ Brussel

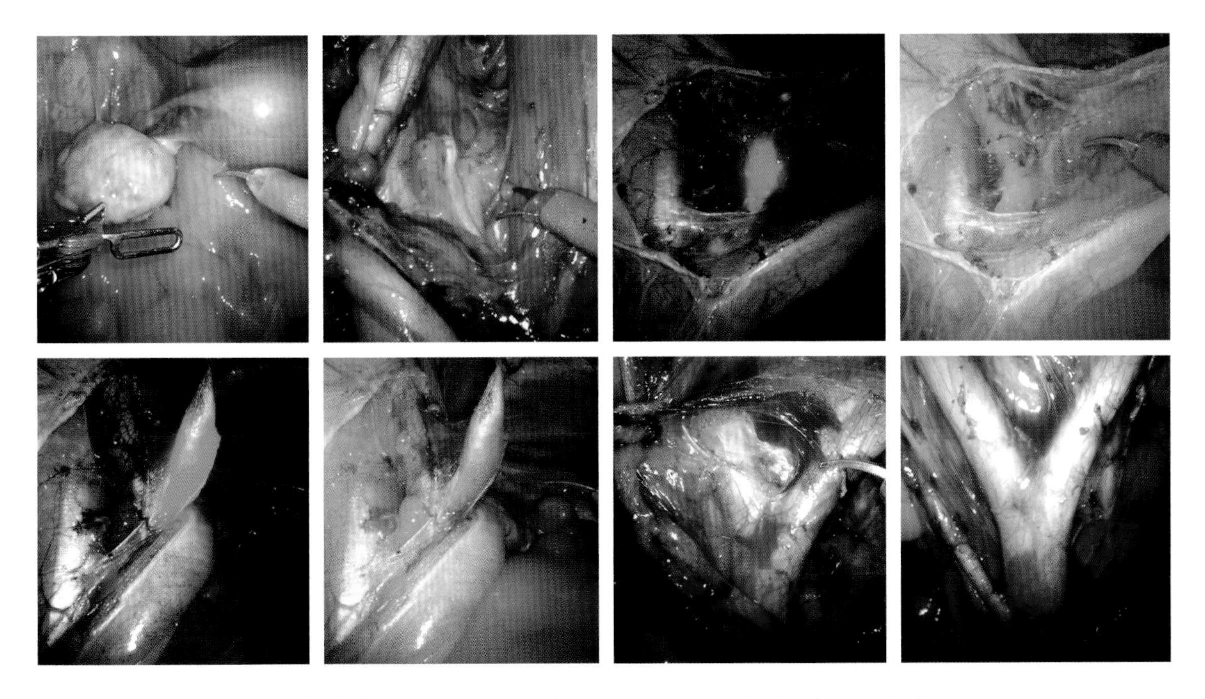

Photographs of robotic gynaecological cancer surgery (lymphadenectomy and sentinel node procedure).
©UZ Brussel

The advantages of robotic surgery

The surgical techniques used are similar to conventional endoscopy – or keyhole surgery – which has been common practice in many procedures and surgical specialisms for about thirty years, so you might be fooled into thinking this is nothing new. However, thanks to the excellent visualisation capabilities and extreme precision of the robot-controlled instruments, the quality of surgical intervention has improved enormously. This means that more procedures, and procedures of increasing complexity and delicacy, can now be performed by endoscopy rather than open surgery and that the benefits of endoscopic surgery – reduced pain, less blood loss, fewer infections, fewer complications, shorter hospital stays, faster recovery and quicker return to work – are available to an increasing number of patients, without any loss of quality or worsening of medical outcomes.

Robotic surgery is currently most commonly used in gynaecology, urology (kidneys, urinary tract, bladder, prostate, etc.), abdominal surgery (oesophagus, stomach and intestines) and thoracic surgery (thoracic cavity). The strongest indications occur in cancer, where most procedures of this type are performed, for example the removal of the prostate or kidney, uterus, colon or rectum, lymph nodes and lung tumours.

Also new is the single-site technique, in which three robotic arms, a camera and two instruments are inserted into the umbilicus (belly button) via a single small incision, meaning that the patient is not left with any visible scar. However, the technique is limited to relatively simple procedures such an ordinary hysterectomy.

Hefty price tag

However, robots like this are very expensive and we have to weigh up whether this investment is justified. In the early days, there was a lot of scepticism as to whether robotic surgery could really offer the expected benefits. The purchase price of these systems varies between 1 and 2.5 million euros depending on the type, configuration and options and an expensive maintenance contract is also needed. In addition, the cost of using a single instrument during an operation is approximately 300 to 400 euros with at least two, but sometimes three or four, being required. Additional surgical accessories and aids are equally expensive and the payouts offered by the Belgian insurance system for this are not sufficient to cover the cost.

Most hospitals are therefore making a financial contribution themselves to benefit patients and improve quality of care. Some, but not all, hospitals pass these extra costs on directly to the patient.

There has also been some degree of scientific controversy: the costs of the system and procedure have to be weighed against the benefits in terms of efficiency and quality of patient care. Most comparative studies indicate that robotic surgery is superior to open surgery in several areas but that, when compared to keyhole surgery in the abdominal cavity, the robotic option is much more expensive but no better.

However, things are changing. Initially, robotic surgery was accused of being very slow and time-consuming compared to conventional procedures, and robots were not used to their full potential in some services. But thanks to their technical advantages, the latest generation of surgical robots can perform the same operation as fast or faster than surgeons using keyhole surgery, and work more efficiently.

Due to the increasing number of indications for this approach and the growing number of specialists using robots, all the systems that have been installed are now being used at much closer to optimal capacity. This is bringing down the cost per operation.

It is always difficult to demonstrate that a new approach is clearly better than an existing one using scientific comparative research and, ultimately, it is practical experience that determines whether or not new techniques are introduced. In this case, the benefits of robotic surgery for both patient and surgeon are clear: surgeons who can operate on a patient effectively in comfortable conditions using optimal instruments achieve better results. This consideration is particularly important for complex cancer operations.

Surgeons are still important

Despite all its merits, robotic surgery is not a solution to the problem of bad surgery: after all, *a fool with a tool is still a fool.* Its many technical advantages mean that it is easier to learn than traditional keyhole surgery, and the learning curve is shorter. But the knowledge, experience and skill of the surgeon behind the robot, as well as those of the surgeon's support team, remain crucial to the medical outcome. The training of young surgeons should therefore begin with basic knowledge of anatomy and conventional surgical techniques, with more complex procedures – if possible using endoscopy and only where appropriate and justified using expensive robotic surgery – being learned at a later stage.

Like other countries, Belgium has a specialised training and research centre for robotic surgery, the ORSI Academy in Melle. Surgeons, as well as surgical assistants and nurses, need to be trained in the special procedures and techniques inherent to robotic surgery. After all, they stand at the operating table by the patient's side and work with the operating surgeon, who sits a little further down at the console, to keep the operation on track. Good teamwork based upon clear agreements and smooth communication is therefore of the utmost importance. This is precisely why telesurgery, in which the surgeon sits much further away from the operating table, may work less well, despite being technically feasible.

As with all aspects of healthcare, quality goes hand in hand with experience. To build up that experience in a specific technique, a surgeon and their team must treat enough patients with a particular condition. The centralisation of rare or complex procedures is therefore highly desirable, but nowadays this provision is scattered throughout Belgium's entire hospital landscape. The creation of hospital networks in Belgium could be an appropriate lever for concentrating such expensive systems and pooling experience and investments.

A known quantity

So, why are so many hospitals investing in expensive robotic systems if these systems are so controversial and difficult to finance? Just like any company in a competitive marketplace, every hospital wants to offer optimal care using the latest and best equipment operated by competent doctors. But the question of whether, given its high costs, robotic surgery offers patients sufficient added value to be introduced into healthcare on a broad scale is one that keeps coming up. After all, these are expensive systems and their additional benefits in clinical terms cannot yet be backed up by hard objective evidence.

This is one of the reasons why the University Hospital of Brussels (UZ Brussel) waited for a relatively long time – until 2015 in fact – before establishing a programme of robotic surgery. Extensive research and visits to national and international centres with years of experience have allowed us to avoid the errors of the past, and robotic surgery in its current form has now gained an undeniable place within the surgical arsenal.

UZ Brussel decided to introduce the technology under controlled conditions, i.e. within a structural programme supervised by an autonomous steering group in which indications, outcomes, developments and training courses are centralised and followed up. A specific innovation budget was made available to finance an agreed number of procedures per year, meaning that patients don't have to make a financial contribution to cover the extra costs. Thanks to this approach, several surgical disciplines at UZ Brussel have quickly become national and even international points of reference.

We've come a long way

Back in 1999, the first commercial surgical robot was a cumbersome, slow and unwieldy machine. By contrast, the fourth generation da Vinci Xi has now evolved into a highly sophisticated, fast, flexible and accurate precision instrument. The system can be used for an increasing number of procedures and for surgical indications for which it would not have been an option in the past.

Despite the monopoly position and huge technical lead of the da Vinci robot, its manufacturer continues to invest in perfecting the system and expanding it to incorporate new, related technology – and the number of systems sold worldwide continues to grow. Very soon, a number of major manufacturers will introduce a new and somewhat similar robotic surgery system that has been approved by US watchdog FDA onto the market, and a wider range of robotic or hybrid keyhole surgery systems will find its way into operating theatres. It is possible that this diversification and the breaking of the monopoly will lead to a general fall in prices and give users – hospitals, doctors and public authorities – greater bargaining power.

And we've still got a long way to go

There's no way back – the technological revolution cannot be reversed. However, supervision remains a crucial aspect of medical and surgical robotics. All are agreed that robots can act as an extension of the surgeon, but cannot replace human beings altogether. Work on the development of robotic systems with greater autonomy that can perform actions such as making an incision or stitching a wound automatically has been ongoing for years, but the more complex the structures, anatomy and changing tissue textures, the more difficult it is to teach a robot how to deal with them. The robot would have to determine the best strategy and technique using extremely objective criteria and based on a range of parameters (such as anatomy and tissue structure) obtained from various sources (real-time surgical images and pre-surgery CT or MRI scans). For the moment, crucial decisions need to be left to the surgeon.

In the very near future, the introduction of virtual or augmented reality, image-guided surgery, semi-automation and highly compact single-port systems is set to radically change the role of the surgeon behind the robot, and that of their operating team.

On the left, the four-armed robot towering above the patient and, on the right, the 3D glasses and levers of the surgeon's console.
©UZ Brussel

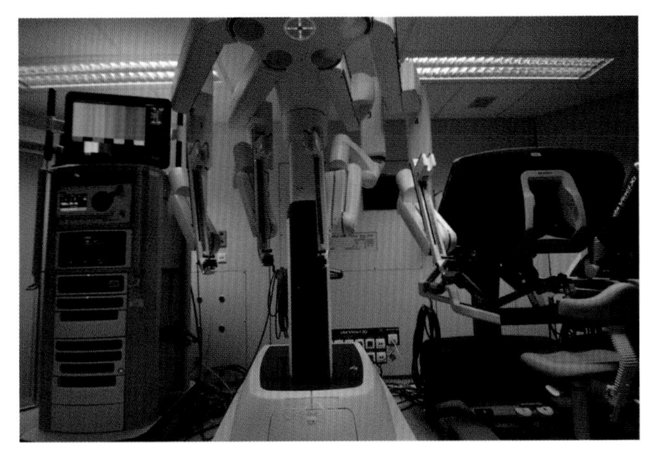

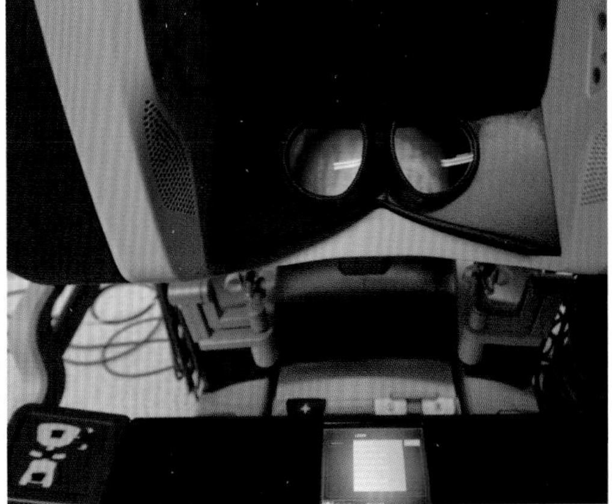

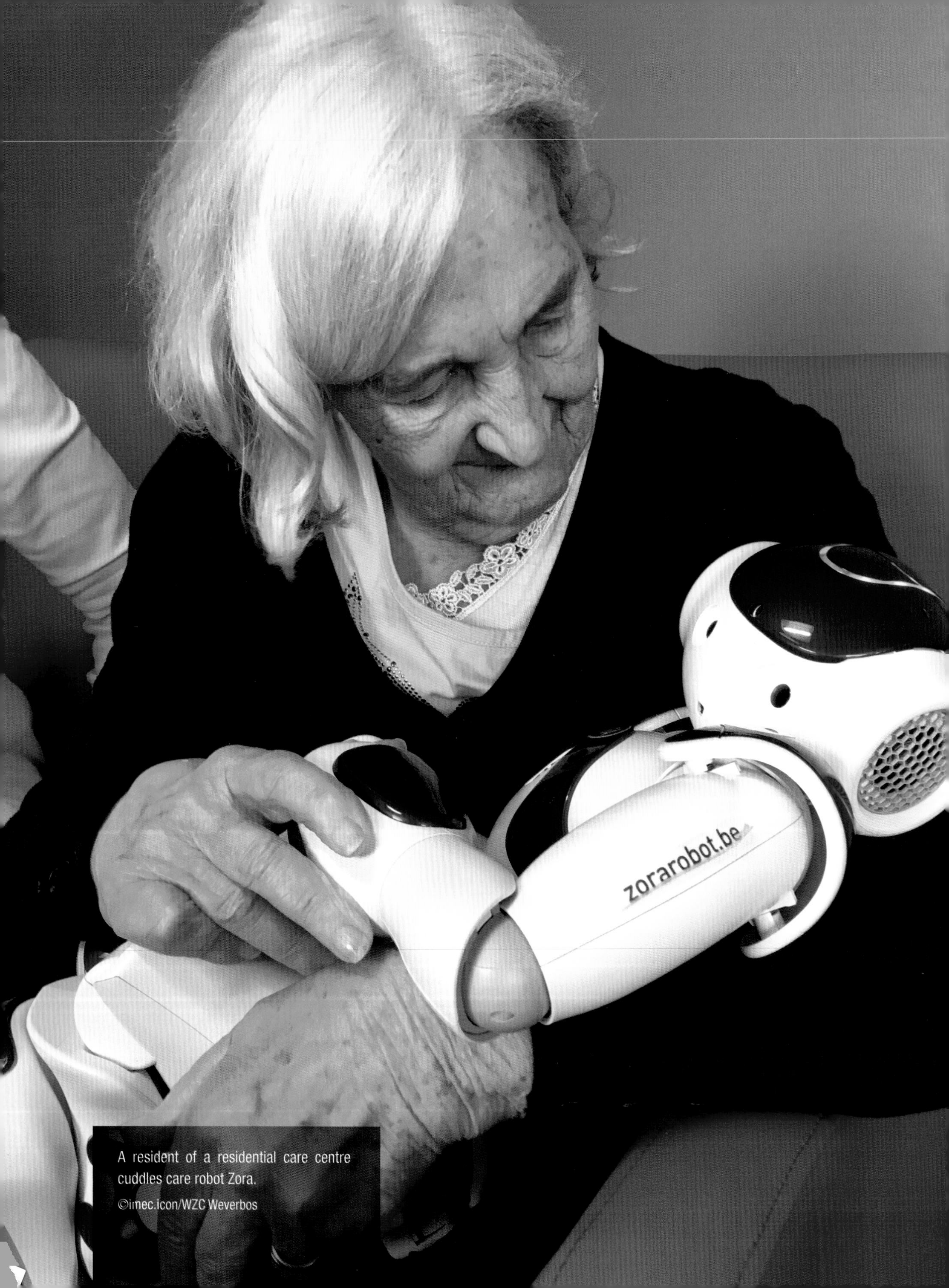

A resident of a residential care centre cuddles care robot Zora.

©imec.icon/WZC Weverbos

CAN A ROBOT PUT YOUR GRANNY TO BED?

By Prof. Dr. Nico De Witte, Prof. Dr. Dirk Lefeber, Shirley Elprama and Prof. Dr. An Jacobs

Most western societies are facing the challenges of an ageing population, with the proportion of older people increasing, and this trend is set to continue over the coming decades. According to a recent study by Statbel, the proportion of people over 67 compared to the number of 18 to 66-year-olds, i.e. the working population, is 26 percent. By 2070, that percentage is set to rise to 39 percent: in other words, four out of ten Belgian residents will be over 67 years old. This means that there will be far fewer people looking after a larger group of older citizens. Estimates for 2060 predict there will be one older person for every two that are still active. The expected increase in the number of older people often goes hand in hand with care needs and chronic conditions.

Many governments are committed to de-institutionalisation as a solution to this problem, with older people only allowed to go into a residential care facility if they are very care dependent. Their only alternative is to stay in their own home, known as *ageing in place*, and the fact that older people say they would prefer to stay at home for as long as possible is welcome news for policy makers.

However, this policy raises questions about who will take care of all these community-dwelling older people. Professional caregivers are already short staffed and a whole generation of nurses is due to retire soon. Because these jobs are not being made financially attractive, the number of graduates entering the sector is scarcely enough to cover the outflow. So it shouldn't come as a surprise that care robots are seen as a way of helping people to remain as independent as possible as they grow older at home.

Robots are no substitute for a nurse

Just a few years ago, no-one had even heard of care robots. Now it looks as if they could be the ultimate answer to expected staffing shortages in the care sector. But we need to rein in our enthusiasm. The fact that some people already have robots around the house in the form of vacuum cleaners or lawnmowers doesn't mean that robots who are able to carry out a wide range of caring tasks will be around any time soon. Taking care of someone is complex and there is more to nursing than the purely technical and functional aspects of the profession: psychological and social factors are also important. Healthcare robots not only need to be able to perform a task, they must also be able to think proactively, empathise, anticipate, evaluate and carry out other duties. By contrast, existing robots are only capable of executing specific tasks.

Social robots

Social assistive robots are robots that can communicate with people in a natural manner, often using speech, emotions and gestures. Their job is to keep people company and carry out simple tasks. For instance, a desk robot is a device with robotic features that can be set on a table or desk. One example – until the company that made him went bankrupt – was the desk robot Jibo. Another example is Billy-Billy, a kind of talking flowerpot from ZoraBots.

Both devices are loaded with technology and can see, listen, speak, help and even learn. You can connect them to the internet, allowing them to read the weather forecast, tell you what time it is or the temperature of the room, and read out messages from friends or family.

All very wonderful and fascinating – the robot can do a lot of interesting things at your request – but as yet no substitute for a real conversation with a family member, friend or neighbour. Moreover, robots generally can't see, listen, speak, help out or learn as well as people can.

The cuddly robot Paro, in the form of a seal, was developed in Japan to reduce anxiety in people with dementia. The seal responds to sound, touch, movement and light, with sensors allowing it to continuously adjust its behaviour depending upon the stimuli it receives. Cuddly robots like this are an attempt by developers to mimic the therapeutic effect that animals have on human beings, bearing in mind that, in Japan, pets are often not allowed in nursing homes and hospitals.

Another good example of this type of robot is Nao. Nao is 57 centimetres tall, weighs 12 kilos and serves as an instigator. He can sing, read out bingo numbers or keep you abreast of the day's news. But Nao can also help you in exercise therapy by standing at the front of the room demonstrating you the movements. This gives the physiotherapist, who would otherwise be the one demonstrating the exercise, the opportunity to provide individual assistance to residents without interrupting the exercise for everyone else. Experiments are also being done into how Nao can be used with children with autism. The robot doesn't work miracles – it is intended as an aid rather than a substitute for a human being. It can demonstrate exercises so that the therapist can focus on the child. Although the robot has small arms and hands, it cannot handle physical tasks such as bringing coffee.

So, we are still a long way from the robot in the film Robot and Frank, in which the protagonist Frank (Langella), who is showing the first signs of Alzheimer's disease, establishes a friendly relationship with a humanoid robot who not only assists him socially, but also performs physical tasks around the house.

Service robots

In addition to the social side of the job, nursing staff also provide a wide range of physical services. The self-propelled robot AethonTug can deliver medicines and fresh linen and dispose of waste – but pressing the button in a lift or using a door handle is a step too far.

It would be enormously helpful – and not only for older people – if robots could fold laundry. Two robots have been developed to do this job, the Laundroid and a reprogrammed cobot by Rethink Robots. These two robots can fold clothes, but they are still very slow, with the Laundroid taking four minutes to fold an item and the Rethink Robot as long as quarter of an hour. In contrast, a human being can perform this task in a few seconds without even thinking about it.

Robear looks like a friendly bear and is nearly as strong. It was developed as an almost autonomous device to help people change position, for example to get them into and out of a bed or wheelchair. It's an ingenious and much-needed concept, as the daily job of lifting and moving patients places a very heavy burden on nursing staff. However, it's important that patients buy into the idea. As it is, nursing staff frequently encounter suspicion or outright refusal when they want to use a hydraulic lift. In response to this, exoskeletons are being developed in order to reduce the physical strain on their bodies during lifting.

Robots with that little bit more

The robots of the future will be integrative devices that can handle a range of different tasks. After all, it would be madness to surround an older person living at home with five different robots: one for company, one to guide them through their daily exercise routine, one for vacuuming, one for folding clothes and one to help them out of their bed or chair. Not only would this be ridiculous, there would also be a expensive price tag for purchasing these devices and for maintenance and repairs, not to mention logistical aspects such as charging. If healthcare robots are really going to conquer the market, they need to be able to take on multiple tasks. And, what's more, they need to be able to empathise, think proactively and recognise emotions. Despite the impressive capabilities of robotics and artificial intelligence, the perfect robotic butler is still a distant dream.

Care from the heart

The debate on care robots can be approached from various different angles. Some people are sceptical and take the view that care robots will never be viable, whilst others worry about declining levels of care, fearing that genuine care with a personal touch – already under pressure due to economic factors – may become a thing of the past. Yet others see opportunities for robots in healthcare, arguing that having robots carry out the routine aspects of the job autonomously allows staff to focus on providing genuine personal care that comes from the heart, something that is all too often neglected.

So, the subject of care robots is one that arouses strong emotions. In this debate, one group is barely involved: the end users themselves. The future acceptability of healthcare robots will depend on the degree to which end users have a say in their development. As yet, there has been little research into the opinions of care users about the use of robots, or into the question of which robots work best for which target groups. It is not inconceivable that patients will see things differently than developers, designers and care providers. Can you imagine giving your 85-year-old grandmother a talking flowerpot as a gift? How do you think she's going to react?

Patients and carers need to be given a say

In an ideal world, the further development of care robots would therefore be a process of co-creation involving all the main partners (both end users and care providers), making it easier for everyone involved to accept the robots. After all, there's no point investing millions in a high-tech robot that can work wonders, but that nobody really wants.

During co-creation sessions with healthcare professionals held as part of imec.icon's WONDER project we learned that end users sometimes have different expectations of robots than was anticipated by developers. For example, they expect them to be much more autonomous when it comes to detecting problems, such as conflicts between residents.

Working in co-creation with healthcare professionals and patients also gives rise to a much more user-friendly end product. The aim is to produce a robot that everyone finds easy to use, but to do this we need to actually ask end users their opinion. All too often, developers get carried away and forget that controls that they find easy to use can be baffling for an ordinary human being. Take Microsoft's Windows 95 operating system, for example, which you could shut down by pressing the start button. This caused problems for lots of people – on an intuitive level, it just seems wrong to press Start to stop something. People approach new things using the mental model they have built up in the past: of course, they can learn to do things differently, but keeping a new system closely aligned with the user's past experience is bound to cut the learning time.

Some of the care robots in use today still need to be directly controlled by the care provider (*puppeteering*), so it's good that developers are giving a lot of thought to how these robots are programmed, for example by making it as easy as possible to set a robot up for an occupational therapy session. Care providers still perceive this as extra work, which they sometimes find hard to take. The development of community platforms where tips, tricks and programs for sessions and exercises can be easily exchanged is thus a very important way of promoting the use of these applications.

A robot in every home

Finally, we need to consider the question of robot integration, not in terms of the various tasks that a single robot should be able to perform, but from the point of view of incorporating the robot into both care processes and people's homes. When it comes to care, one thing is clear: robots that cannot be easily integrated into the care process will quickly come to be seen as deadweight and be relegated to the sidelines. Besides this, the context is also important. It's likely that older people will be the first to encounter care robots, and will have the most to do with them, probably in their own homes. However, every home is different, and they are often not adapted to their occupants' changing needs and requirements.

In the WONDER project, we found that some older people liked the robot Nao, while others wanted nothing to do with it. Anyone who has ever given advice on housing modifications for fall prevention will know that every home has its own challenges and problems. It will therefore be important to develop care robots that can be used across the board and are not restricted by living conditions. But until these much smarter robots are a reality, we need to gain a good understanding of the circumstances under which the existing solutions can be usefully deployed and to who they add value.

The challenges for healthcare robots are therefore very great. A huge amount of progress has already been made and technological developments will undoubtedly give rise to models with ground-breaking features in the future. It seems that symbiosis between developers and end users is the missing link enabling care robots to make the final breakthrough. By working together, we can look forward to an exciting future.

People often find caressing and cuddling animals to be therapeutic, but space and hygiene issues can rule out pets. The robotic seal Paro, developed in Japan, was designed to bridge this gap.

©AIST

LAW AND TAXATION

You can ask the smart loudspeaker Alexa to start the irobot vacuum cleaner.
©irobot

WILL HOUSEHOLD ROBOTS WASH YOUR DIRTY LINEN IN PUBLIC?

By Dr. Rob Heyman, Prof. Dr. Jo Pierson and Prof. Dr. Paul De Hert

Won't robots violate our privacy? For the moment, this is still largely a matter of conjecture, which is why we are engaging in this process of informed speculation. We are going to do this by examining how privacy has already been affected by existing data-driven technologies such as social media, the internet and computers. This shows us that there are two points for consideration. Firstly, transparency: what do we know about the way robots process our data? And, secondly, control: how do we control what robots get to know about us?

What we can learn about privacy from social media and search engines

Data-driven technology collects all sorts of data to improve its services or offer new ones. For example, Google's parent company, Alphabet Inc., offers free email, a search engine, office applications and navigation. We know Facebook Inc. from Facebook, Facebook Messenger, WhatsApp and Instagram.

Behind those services, which are often free, lies a strategy. The aim is to make these services an important – and preferably indispensable – part of our daily lives as quickly as possible. The more you use one of these services, the harder it is to leave it. The companies in question benefit in two ways. First of all, you provide them with more and better information, enabling them to tailor their services even more accurately in relation to their competitors. And secondly, people become dependent upon their services, making it increasingly difficult for them to leave.

In the past, this dependence has on several occasions been exploited to impose things that end users had not explicitly agreed to, in particular the reuse of data for new services or by other organisations.

Domestic robots, which we will come to shortly, will also collect large amounts of data. Some of the data giants of the internet are making the leap to robots in the home so it's possible – but again we are only speculating – that those companies are planning the same thing here.

Social media or online platforms rely on three mechanisms. They collect all kinds of personal and other data on a massive scale, often without people realising they are doing it: data that people provide voluntarily (name, photos, etc.), observed data (such as cookies that monitor your surfing behaviour) and derived data (such as your creditworthiness profile). This data is then sold, which often means it is used to generate more highly targeted advertising. Finally, the data is used to control what users see and don't see, and what they can do. These mechanisms can undermine privacy and other values.

Jibo has been referred to as the first social robot for the home, but this also raises privacy concerns – for example due to the collection, processing and storage of information from its sensors.

©Jibo

The antics of Alexa

The example of Amazon's Alexa is very telling (Alexa is actually an Amazon personal assistant that works through devices such as Amazon Echo, but you could call it a kind of robot). Alexa is constantly listening for commands like 'Alexa, do X.' At times, however, she misunderstands a command and divulges information that another family member might prefer not to share. For example, there is the case of a young boy who asks Alexa to play a song, to which Alexa responds by suggesting the child listen to a porn station. The boy's father, whose search history Alexa used by accident, has been unmasked and can only shout 'Alexa stop!' from the background.

And then there's Danielle's story. In May 2017, she was contacted by a colleague of her husband who had received a message from Amazon containing a recording of a conversation at Danielle's home. Since then, all the Amazon Echo devices in Danielle's family have been switched off. It is still not clear exactly what happened. Amazon say it is Danielle's fault – Echo had asked permission. Danielle denies that, but Amazon is standing firm, saying that the device twice asked permission to send the conversation to the colleague. Was Danielle lawfully recorded or did Amazon make a mistake? No hard evidence has been provided.

In these examples, Alexa accidentally made information from a domestic context accessible to others. Social media researchers say this is nothing new. Pictures on Instagram or Facebook or even your own search history can also reveal this kind of information. One solution is to 'train' Alexa better or for users to check what access domestic robots like Alexa have and then, if necessary, either erase information or make it inaccessible.

Robots know more about you than you think

The threat to privacy posed by domestic robots goes beyond the unwanted disclosure of existing information.

Aldeberan Robotics has a robot, Nao, who asks annoying questions about your private life. Nao uses facial recognition: as soon as it recognises your face it asks questions about things it finds about you online. For example, it might find out where you ate yesterday and ask what it was like at that particular restaurant. Nao is part of a project called *Humans and Robots in Public Spaces* that is investigating the way in which people will interact with robots. In the future, robots will be able to see and sense much more, which will enable them to interact with us in a much richer way, but will also make them much more invasive. Hence the experiment.

As we mentioned above, robots are a data-driven technology. They collect data via sensors or retrieve information that we have made available online. People can collect that kind of public data, too, but it feels even creepier when robots do it. The example of Nao shows that we need to be critical of the new unknown in relation to our everyday lives. Nao is still quite innocent, because it cannot get at the information that Apple, Google and Facebook currently have access to. Personal assistants such as Siri and Google Now know where you are and what you are looking for at all times, and have a good idea of your normal daily routine.

Should we be wary of robots? Will people hide information, as some people put stickers over their laptop's webcam because they are afraid someone might be watching them? The sticker is the result of powerlessness and lack of trust. People who cover up their laptop camera do so because they don't have any way of checking that there really is no-one watching. The level of trust in the companies behind domestic robots is not particularly high, either. Companies like Amazon and also iRobot – the producer of the popular Roomba vacuuming robot – are vague about what happens to our data (or what will happen to it in the future).

The Clean MapTM lets you see where the iRobot vacuum cleaner has cleaned, but this has given rise to privacy issues when the company looked into selling those floor plans.

©irobot

Roomba is an electronic vacuum cleaner with a low-resolution black and white camera, touch sensors and infrared sensors. The robot's privacy disclaimer states that the company may, under various conditions, resell your data.

The boss of iRobot, Colin Angle, has Roomba's maps of peoples' homes sent to iRobot. In the first instance, these maps are used to calculate more efficient routes, because Roomba cannot do that for itself. But Angle felt that these maps could also be interesting to suppliers from the *Internet of things*, such as Google and Apple. Angle talked to the press about selling data, resulting in a wave of criticism. Two days later he took back his words, saying that he wouldn't sell the data, but he would share it if the user wanted.

So not only is it difficult to know what a company behind a robot is doing with the data it collects, it is also anyone's guess what a company might do with this data in the future.

In privacy research, this phenomenon is known as *function creep*. As soon as a system collects data, new purposes or functions for that information are constantly being spotted and applied. Companies find these new data applications irresistible. After all, they have the data already so every new feature they can come up with is pure profit.

Personality inspires confidence

We know that people love to assign a personality to their robots and personal assistants. Roombas, for example, are often given names or treated like a pet. You can also personalise Siri and Google Now if they don't already have a personality of their own. When we personify something, the whole structure behind that 'person' remains hidden to us because it takes place in the background or is invisible. This phenomenon goes far beyond robots, but it is particularly relevant here.

Professor Mireille Hildebrandt warns that that personification could be disastrous, because it could make people trust robots in a way that could be risky. That danger certainly exists. Children, for example, are often prepared to tell bots some really weird and wonderful things just because they look like their favourite TV heroes. And, on Tinder, bots are able to ensnare gullible people who believe they are talking to a human being.

Robots are (still) dumber than you think

Robots in the living room are dumb, but just smart enough to map out their surroundings. It is a computer connected to the robot that does the thinking to determine whether an instruction has been given or produce a clear map of a room so that it can be vacuumed properly. The behaviour of the robots connected to a computer can thus suddenly change if the smarter computer makes an adjustment. Since we tend to personify robots, we find this difficult to judge and it is something we haven't yet learned to anticipate: after all, your washing machine doesn't suddenly start behaving differently because it has received an update. This same personification-driven invisibility also means consumers tend not realise when a certain data flow is being passed on to another company. As a result, another event along the lines of the Cambridge Analytica scandal should come as no great surprise.

The other consequence of the personification of robots is that we tend to believe it is the robots doing all the work, rather than human beings. Robots are sometimes delivered before the smart computer behind them is finished, in which case it may actually be people pulling the strings instead of the computer. This is known as the Wizard of Oz technique.

The 'Invisible Boyfriend' is not a robot but a digital service that allows you to receive romantic messages and voicemails from a simulated friend for a monthly fee. The company behind Invisible Boyfriend wanted to provide a fully automated service, but its technology was not yet advanced enough for this. So it hired human beings to send out the loving messages. The problem was that not everyone understood how the system worked, meaning that many people were blithely chatting away in the mistaken belief they were conversing with a computer.

Recommendations on home robots and privacy

Do the anecdotes cited above allow us to make any predictions about privacy and domestic robots? We think so. Many of the companies that threaten our privacy online are the same ones as those attempting to do the same thing using robots. In that sense you can see all these anecdotes as the tip of an iceberg: it's a given that even more similar things are going on under the surface. And that's our biggest challenge: there is an enormous lack of transparency. It's unclear which parties, people and computers are watching us now, and will be watching us in the future.

So, how can we increase transparency? Here are a few recommendations:

- Where possible, sensors should be designed so that it's clear what they are and when they are on.
- Every robot should come with an information file that allows people to see at a glance what's happening to their data.
- Robots, unlike apps and software, should not simply be allowed to change where they send information or what they do with that information after an update. Since this is a complex matter, it would be best to have a team of experts monitoring privacy.

Robots can help here. We already use robots, or rather programs, to control our privacy preferences in the form of apps and browser plug-ins. These applications do two things: they check what data is being exchanged and they stop certain parties by blocking their access. Robots don't mind doing repetitive jobs that people would prefer not to, so robots are welcome if they are working on our behalf to protect our data from unauthorised access.

But privacy robots alone are not enough. Large companies use their market position to impose choices along the lines of 'agree to the new ways we are going to reuse your data or leave'. And leaving is difficult because these services are embedded into our daily lives. The same thing may happen with robots. Suppose we buy a Roomba that does not provide data to, say, Apple at the time of purchase. But if you don't want to go back to vacuuming your house yourself, you will have to agree to new conditions or your vacuum cleaner will stop working.

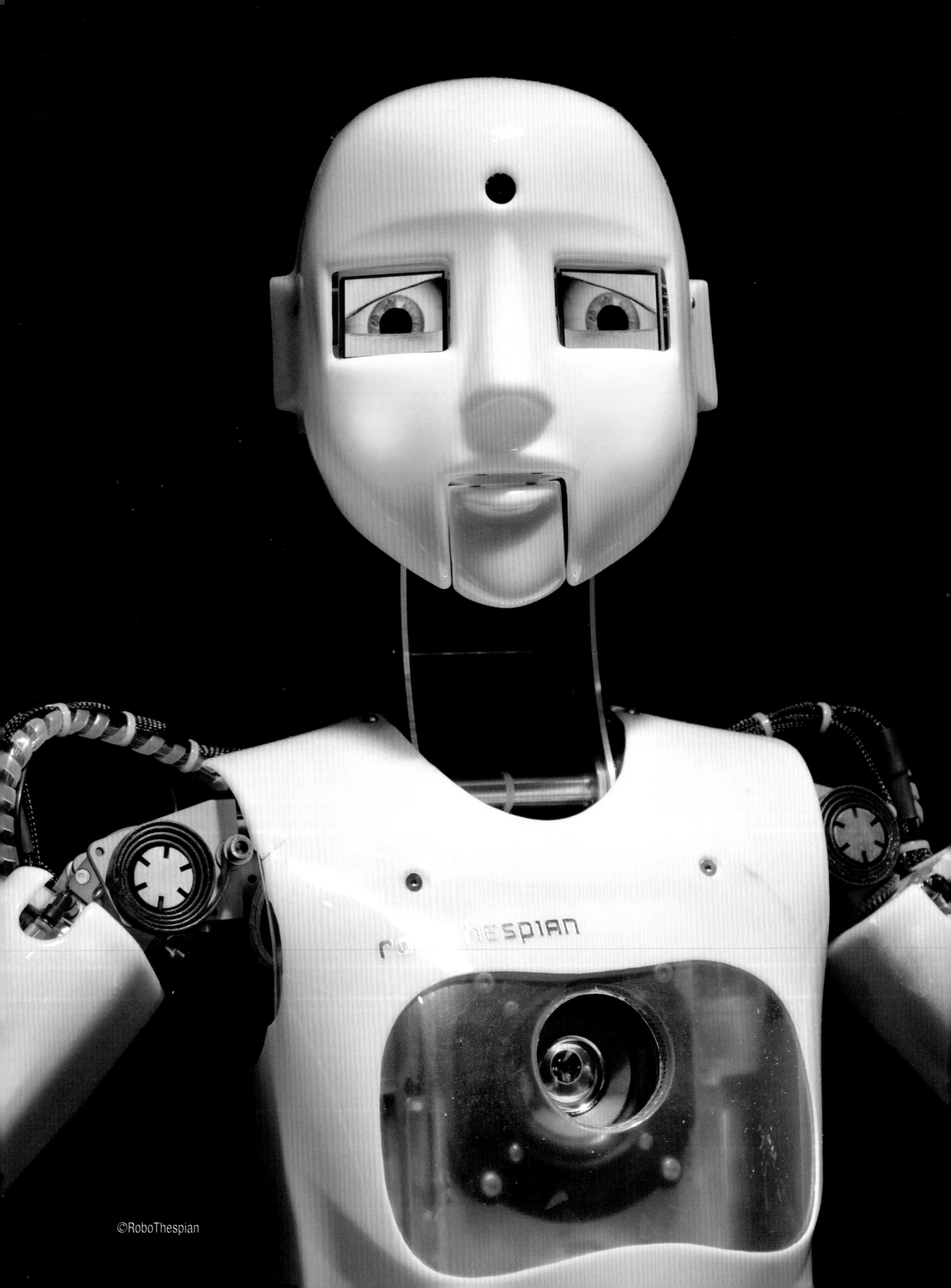

©RoboThespian

WHO'S TO BLAME – THE HUMAN OR THE ROBOT?

By Prof. Dr. Daniel De Wolf

The year is 1386, the location Falaise, France. A baby is found horribly mutilated and dies of its injuries. The finger of suspicion is soon pointed at an unfortunate pig that witnesses saw entering the house. The pig is sentenced to death by hanging. It's a true story.

R. Chambers, The book of days: a miscellany of popular antiquities, 1864.

The year is 2035. Robots are part of our everyday lives, an omnipresent servant to make things easier for us at home, at work or on the road. Del Spooner is a special policeman whose job it is to supervise the robots and investigate any incidents. When an inventor is found murdered all hell breaks loose. Not only does Spooner suspect, much against the tide of prevailing opinion, that a robot had a hand in the crime, he soon finds out that there's more to the case than meets the eye and that humanity itself is under threat from a robotic conspiracy. This fictional narrative is the plot of the popular film, *I, Robot.* But could it ever become reality?

Of pigs and robots

Lawyers sometimes have to grapple with some peculiar problems. The stories above illustrate how the same legal questions can crop up time and time again even in very different times. The context may be different, but the underlying issues are the same. Can you see the connection? No? Let's start with the past.

From the Middle Ages to the eighteenth century, people have wrestled with the question of whether animals could be held criminally responsible for crimes. A range of sources indicates that animals regularly appeared in court as defendants in the Middle Ages. And these weren't minor charges: animals were accused of rape or murder, with the trial of the pig described above being just one of many.

Today, animals can never be held criminally liable as they are considered to be 'things', with criminal liability being restricted to adult human beings.

Medieval lawyers had not yet reached this conclusion and held heated debates on questions such as whether animals have souls and whether they can be classed as people. They reasoned that if so there would be no reason not to punish the animal, with the penalties imposed often being even stricter than those for human beings. It was considered morally reprehensible for an animal to commit a horrible act and punishments included the death penalty or exile to galley ships. Some lawyers – or their predecessors – even specialised in the defence of these unfortunate animals although, strangely enough, this was only permitted in the ecclesiastical courts. In rare cases the animal was acquitted and a human being found guilty of murder or fornication.

The development of robots and artificial intelligence is giving us cause to open up these questions once again: can a robot be punished? Can it appear in court accused of a crime? What penalties should it face?

The 2004 film *I, Robot,* sketches out a fictitious technological world. The story is about the danger we face when robots are an omnipresent force in our lives, and about the illusion that we can stay in control of them. Ultimately, the artificial intelligence in the film sees man as a threat, but fortunately there is a police force supervising the robots and one heroic officer is able to save humanity. The interesting point here is the thought that robots can commit crimes and that a special kind of police force (made up of human beings) is needed to control them. Let's see if we can find some answers to these questions and investigate the issue of whether robots can bear criminal liability.

The case of the self-driving car

When it comes to human beings, it's simple. A criminal is someone who performs an act that is forbidden by law or who fails to perform an act required by law. The action or omission is called the crime, the person the perpetrator, and the perpetrator is duly punished. This is the concept of criminal responsibility. Normally, only human beings are regarded as 'persons' under criminal law, but sometimes the question also arises of whether non-humans should be able to bear criminal responsibility.

As we have seen, this question arose for animals, and was rejected. But it has been a fairly recent development in Belgium that allowed legal entities such as enterprises and companies to bear criminal responsibility that is separate from the responsibility of the persons that work in or run them. So it's clear that we need to ask ourselves whether robots can be held criminally liable. And, if not, who should we hold responsible for actions performed by a robot? The case of self-driving cars illustrates a number of points.

The question of whether a robot can be considered to be a person is discussed elsewhere in this book, so the next question must be whether the robot exhibits the characteristics of a human being, and whether it is desirable to equate the robot with a human being and assign it a 'personality'. Criminal law does not require a 'person' to fulfil any particular characteristics. Let us turn for inspiration to the situation that exists under Belgian criminal law, which states that to be considered a legal person an entity need only perform actions for itself, or through a human being, and have a limited form of independent will, which may take the form of a particular policy that is distinct from the will of a human being. It must also act in line with that policy, otherwise it cannot be held responsible for its behaviour. In short, it's perfectly possible for a robot to be held criminally responsible.

But, of course, this does not address the question of whether this is what we want, and there are a few issues that need to be considered. For example, it would be rather strange to impose obligations on robots without also giving them rights, but granting rights to robots might be taking things a little too far. Moreover, unless we create shared responsibilities, the creators (the manufacturers and software developers) may leave the robots to shoulder the blame while they themselves get off scot-free. After all, if someone releases a dangerous dog and it bites a child, we don't put all the blame on the dog, and let the human being off the hook – allowing people to shirk their responsibilities is not a good approach. So independent responsibility may not be the way forward.

Things get even more complicated if we assume that the robot, computer or machine is not in full control of its actions, but works in tandem with a human being or legal person. In such situations, the criminal responsibility is not clear. For example, fully autonomous cars without a driver are not currently allowed on the road: as yet only semi-autonomous cars that still need a human being behind the wheel (several manufacturers are testing such vehicles), and systems that assist the driver (e.g. automatic braking when an obstacle is encountered) are permitted. At present, road traffic legislation only applies to the driver, i.e. a human being. It may be that the rulebook will have to be completely rewritten to allow for the introduction of self-driving cars.

Certain manufacturers are terrified of being held liable for accidents. And another problem is that manufacturers are currently only responsible for the product as delivered. If the machine has a self-learning element to it, can we hold the original manufacturer responsible for the mistakes that result if this process goes awry?

Moreover, if human beings are still responsible for some of the decisions – and can make mistakes – it's not exactly fair to punish the manufacturer when accidents occur. Things are even more complicated if a human being fails to intervene despite being able to foresee a dangerous situation arising. Can we hold a human responsible for incidents when the software is so enormously complex?

In addition, it will not always be clear that the machine is not working properly until it is too late. Criminal law is more flexible in such cases than civil law, because joint errors give rise to responsibility for all perpetrators, regardless of their share of the blame – in the event of a car accident, it's entirely possible for both the manufacturer and the driver to be held liable. Moreover, for the purposes of criminal law, the manufacturer will not be able to choose between the safety of the occupant of the vehicle or that of other road users, including cyclists and pedestrians. Both need to be guaranteed.

A robot with a lawyer

Suppose we assume that the robot is entitled to assistance from a lawyer. After all, we are saying that it is responsible, and since we are talking about a criminal act, the robot is liable to be punished. Much will depend on the technology – will the robot be able to defend itself or not? If it can't we are dealing with a similar situation to that of the pig that was brought before the ecclesiastical courts in the Middle Ages. We can leave its defence either to other robots or to specialised (human) lawyers, but the robot deserves a defence every bit as much as a human being does.

Prejudice will certainly play a role. In today's world of DNA and forensic evidence, judges probably would not consider the fact that an animal has entered a house sufficient to convict it for murder. But with regards to evidence we still see judges jumping from unknown facts to conclusions.This is also a high-tech debate that throws up all kinds of technical questions, requiring us to call upon experts.

Perhaps robots need to have a 'black box' like that found in planes to record all their actions and a range of technical parameters. A system like this already exists for self-driving vehicles. As already mentioned, much will depend on technological progress, but we can be sure that there will be many new problems to be solved.

Can you throw a robot into prison?

Punishment is a highly subjective topic that is very much influenced by the values that hold sway in society. All we can do here is mention a few basic principles.

Firstly, we can ask ourselves whether the penalties for robots should be the same as those for human beings. As mentioned above, animals used to receive more severe punishments than people and, today, there are different rules for humans and legal entities. For example, legal entities cannot be given a prison sentence, but they can be dissolved – a kind of death penalty. Financial penalties are another option, with fines being much higher for legal entities than for individuals. But it would seem strange to impose more severe punishments upon robots than those for humans – as was the case for animals in the Middle Ages. Ultimately, it is the beliefs that prevail within society that will settle this question.

A second question is what type of penalties are appropriate. We can approach this question from two different angles. According to French philosopher Michel Foucault, penalties should impact on the property or value that is most important to robots. Citizens in the nineteenth century prized their freedom extremely highly, which is why, according to Foucault, prison sentences are such a common punishment. Appropriate penalties for a robot may include turning it off, resetting it to factory defaults or deleting its memory.

An alternative approach is based upon what we hope to achieve by the punishment. Nowadays the focus has moved away from revenge and towards prevention and social reintegration, with the aim being to enable offenders to function normally in society after their sentence. This principle could be applied to robots by editing their coding. Rehabilitation is a very popular aim of punishment, with community service being one example and this could be applied to robots by making them work for humans. However, this will not always be possible – a robot designed to work in an assembly hall will find it difficult to sweep up dirt. And, as with community service sentences for human beings, some will fear that this is not enough of a punishment. Another option might be a ban on running a business or practising a profession.

Just as the punishment of animals in the Middle Ages raised many questions and stirred up much debate, so too will the prosecution and punishment of robots. Criminal responsibility for robots does not seem to be out of the question, but prosecution and punishment opens up a whole range of issues. In any case, the parallels from history are clear and there is still much to discuss.

Domi-copter delivers a freshly baked pizza.
©Domino's UK

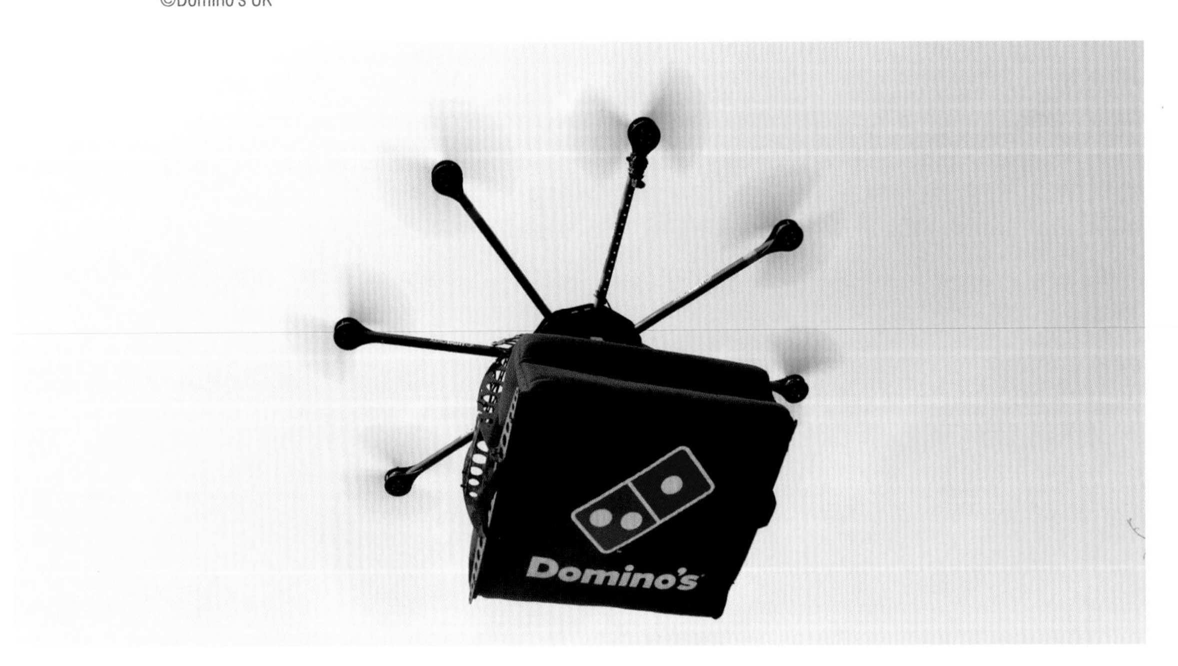

DO ROBOTS HAVE RIGHTS AND OBLIGATIONS?

By Prof. Dr. Mireille Hildebrandt

Suppose we outsource certain aspects of care for the elderly to robots that are able to lift, wash, feed and look after us, and can also have a nice chat and put us at ease in spite of our dependent situation. They can also check whether we are taking our medication at the proper intervals, and make sure we haven't had a nasty fall or become confused. And they can access a wide range of databases and search engines to help with problem diagnosis, alert or consult doctors and nursing staff, and notify our insurance company.

Suppose you have such a robot in your house, which you call Henry and get along pretty well with. What if you have a fall because the robot is not set up correctly (like the self-driving Uber car that mistook a pedestrian in the road ahead for a plastic bag)? What if your insurance refuses to honour its commitments unless you make all the behavioural data collected by the robot freely available to a whole range of third parties to whom the insurance company resells your data? What if you're tired of the robot sharing all your diagnoses with your family without your permission? What if it turns out that your healthcare robot is using your laptop to hack into your friends' bank accounts for its own gain? Who do you turn to for compensation? Can you have the robot hauled off by the police? Can you order a new robot or require the installation of new, less intrusive software? And, if you do, can your healthcare robot appeal against these measures and take legal action because the new software violates its rights as a robot?

Robots are not all alike

Is a cleverly put together puppet a robot? Should we grant such a robot rights or hold it liable? The robots that hit the headlines often look like human beings – for example Hiroshi Ishiguro's robotic twin and Sophia, who was granted citizenship rights in Saudi Arabia. But experience shows that the more closely a robot resembles a human being, the less likely it is to be intelligent or able to act on its own initiative. These robots are more like ventriloquists' dummies: not very smart and almost completely controlled by their creator or user.

When we ask whether a robot can acquire legal personhood we are talking about robots that can, to a certain extent, make independent decisions that are not entirely predictable. Not to the robot's designer or creator, not to the person who puts it onto the market, not to the person who deploys it to provide services, and not to the end user. If Henry is controlled remotely by a nurse, it would be ridiculous to hold it liable for any damage he causes. But if Henry learns new behaviour through interacting with you, gives advice or intervenes without its maker or seller having much influence, we need to ask whether it makes more sense to hold Henry liable than trying to impute liability on others. This argument is often put forward as a solution for situations when no-one wants to take the blame and everyone is saying 'that's not my responsibility, how was I to know Henry would cause all that trouble'.

Limited liability

In 2017, the European Parliament adopted a resolution on rules for smart robotics under private law. As in this book, the term robotics is used in a broad sense to cover self-driving vehicles, care robots, medical robotics (surgery and rehabilitation), cyborgs (people with artificial organs, implants, prostheses, improved memory and concentration), education and employment (new skills are needed in data-driven environments). The resolution calls on the European Commission to draw up legislation governing the private liability of the creators and users of smart robotics, be it strict liability or a reversal of the burden of proof. Compensation for damages can, for example, also be provided by compulsory insurance or a fund.

The European Parliament has also proposed that a special legal status be considered for autonomous robots, in a move that would define them as 'electronic persons' that can be independently held liable for the damage they cause. If we grant such *e-persons* legal personality, they will be able to conclude contracts on their own account and be held liable in the event of damage. This would require that they be certified – not every 'doll with electronics' would be eligible – and have access to their own financial resources, otherwise there would be no possibility of redress.

This legislation would give anyone who has a contract with such a certified robot some degree of protection against negligence or other harm. For example, if Henry is registered as an e-person and has a bank account, you could sue it for damages.

However, there are two main objections to separate a legal status for robots. Firstly, this could be a way for developers or sellers of robots to dodge their liability. And, secondly, this status is based on a misunderstanding about the nature of robots and artificial intelligence.

We need to ask ourselves what problem we are trying to solve by introducing robot rights. In any event, the aim cannot be to exclude the liability of developers, vendors and users under the heading of trying to avoid the stifling of innovation. Innovation is not an end in itself: we need to encourage the kind of innovation that makes a responsible contribution to solving social problems without shifting the risks and creating a whole range of new problems for future generations.

Anyone looking to profit from developing, selling or using robotics should be under an obligation to minimise the damage and risks involved. Making the robot itself liable would only be worthwhile if this could help to prevent such damage and risks from arising. If Henry's own liability leads to developers or retailers washing their hands of the consequences.

The second objection relates to what robots and other applications of artificial intelligence are capable of. Suppose you don't like Henry's behaviour and you request an update or even a different operating system – for example a version that integrates *privacy by design*. Can Henry do anything about that? Can it appeal under some kind of robot law? Is there perhaps a Universal Declaration of Robot Rights that is comparable to the 1948 Universal Declaration of Human Rights?

Universal Declaration of Robot Rights

The answer is no, at present there is no such declaration. But it's still a good idea to raise the question of universal robot rights, precisely because it focuses our attention on human rights. After all: why do we have such rights, and who is this 'we' in any case? First of all, we have to realise that the question of who is granted human or robot rights is a political question, which needs to be answered by the democratic legislator. After all, rights are an artificial construct and not some kind of natural law. They depend on a carefully constructed network of powers that keep each other in check.

Fundamental rights can be seen as the underlying grammar of a society that – as set out by the well-known legal philosopher Ronald Dworkin – demands that governments act on the basis of equal respect and concern for each individual. This equal respect gives rise to the 'one person, one vote' logic of democracy. And it also means that the power of the majority must be limited: in developing and implementing its policy, the government must not completely dismiss individuals as subordinate to the will of the majority.

That makes democracy a difficult but interesting phenomenon. It is not the sum of individual political preferences, taken as a given, but a temporary exercise of power within the framework of the general interest, which needs to be continuously reinvented and re-established. By definition, the exercise of democratic power is temporary and representative because, once in power, the majority must rule also on behalf of existing minorities. The principle of equal respect and concern for to each individual thus underlies both democracy and the rule of law.

When we consider robot rights, therefore, we need to ask ourselves whether we want to make them part of the 'social contract'. Should the government grant robots the same respect and concern as people? It may help us to consider why we do not give animals and plants universal rights.

Animals and plants are living organisms that are sensitive to their environment in a way that robots are currently not, but they are not capable of reflection or conscious anticipation. Holding them responsible for their behaviour would therefore be a tricky business. To have rights you must be able to take responsibility for your behaviour. People therefore have obligations towards animals, but animals cannot be said to exercise any rights. This is because they could only do this through human representation. That is not a problem in itself: after all, companies need to be represented in court. But in that case it is clear who can represent a company, and this is not the case for animals. Who should represent an animal: its owner or an animal rights organization?

Autonomous robots can anticipate us and adapt their behaviour to ours, but they have no awareness, and certainly no self-awareness. Robots, in a sense, have nothing to lose: even though they may be quite good at simulating a limited range of schematised emotions they do not experience pain, suffering, outrage, fear, happiness, pleasure, beauty or renewal. This also raises the question of whether people have obligations towards robots – regardless of whether robots can legally enforce those obligations. Can you put Henry out with the trash or break it apart with a hammer if you find out that it passed on your bank details to a botnet while chatting to you?

Will robots learn to think?

The European Parliament resolution states that 'ultimately there is a possibility that in the long-term, AI could surpass human intellectual capacity'. As mentioned above, robots do not yet have awareness, let alone self-awareness. If they do ever attain this, that could be a reason to grant robot rights to autonomous robots that develop, or have the capacity to develop, a form of self-awareness.

However, we need to bear in mind that robotic intelligence and autonomy is not comparable to that of humans, because it is based entirely upon computation: it's all about logic, statistics and mathematics. To the extent that robots generate meaning, this meaning is attributed by humans. Human thinking is all about meaning and there are no indications that we can design computer systems that are capable of attributing meaning to the symbols they manipulate. Computational thinking is what Hannah Arendt called 'calculating thought'. There's nothing wrong with that – in fact, it's very useful – but it's very different from the way we relate to the world.

Robots can therefore, if the legislator so wishes, be granted legal personality. This only makes sense if and to the extent that it protects those who suffer damage or other harm. It should not be used to provide an escape route for those responsible for the damage, or to protect machines that are good at counting but have no idea of anything else.

Will our rights be eroded?

Philosophising about whether robots should be granted legal personality is interesting, but it can divert attention from more pressing legal issues. The integration of robots into our physical and institutional environments is radically changing those settings. We need to ask ourselves whether, and to what extent, this erodes existing human rights.

In robotics there is the concept of an 'envelope'. This is the area immediately surrounding a robot and is often designed and developed at the same time as the robot itself. The robot's envelope is there to protect nearby people from physical hazards and other harm and represents an attempt to control the environment to minimise the risk of injury. In practice, this is often done by creating separate spaces. A robot's envelope should also increase its productivity, efficiency and effectiveness by making the environment more predictable, manipulable or manageable.

Self-driving cars are safest and can travel fastest when they encounter as few vulnerable people (e.g. pedestrians) as possible and are able to predict the behaviour of the road surface, verge and other road users (e.g. slippery road surfaces, soft verges, traffic jams or unexpected manoeuvres by other cars). Public roads and the entire urban infrastructure will therefore have to be comprehensively redesigned to incorporate sensors in the road surface, separate routes for vehicles and pedestrians, and charging points for electric cars. Similarly, because healthcare robots need to constantly anticipate how their interactions will affect vulnerable patients or elderly people, their environment needs to be equipped with sensors (cameras and devices to detect sound, movement, temperature and smell) that map and dynamically predict behaviour, so that the robot can optimally adapt its own behaviour in real-time. In short, in contrast to industrial robotics, the integration of robotics into everyday life will require far-reaching adaptations of our daily environment.

We are already used to these changes in online settings, where end users function as the environments for data-driven systems. In fact, in many cases, end users actually constitute the envelope of these systems, enabling them to run smoothly thanks to a continuous flow of behavioural data. This gives rise to problems with privacy and data protection, but also to all kinds of difficult-to-detect discrimination, and may even spell the end of the presumption of innocence. Henry exists by virtue of a continuous flow of data that not only maps its physical environment, but also records and calculates your behaviour in detail.

REEM is a full-size humanoid service robot.
©PAL Robotics

The intermediate goal of social networks, search engines, smart cities and other platforms is to influence users in such a way that they become predictable and 'controllable'. Underlying this is the desire to make a profit or serve other private or public interests, such as aiding traffic flow or supporting the elderly. The problem with this intermediate goal is that it is the users who live with and inside such systems who are, in a way, being redesigned.

Henry (and all the autonomous systems that make Henry possible) is reprogramming you to provide it with the predictive power it needs. This means that perhaps the most pressing legal problems relating to robots are the protection of our fundamental rigths as we become increasingly dependent on robots and other autonomous systems.

The Kuka iiwa collaborative robot pours a glass of beer.

TAXING THE ROBOTS!

Prof. Dr. Michel Maus

In this chapter, Prof. Dr Michel Maus examines the effects that digitization and robotization will have on the labour market, and the fiscal and parafiscal consequences of this evolution. Because, if the technological revolution is indeed going to cause job losses, this will result in less tax income for the government and less money for social security. We therefore need to have the courage to pose the following question: shouldn't we subject robots to a separate tax system so that we can continue to maintain our welfare state?

Brave New World

Have you noticed, dear Reader, that the world around us is changing at an alarming pace? Over the past few years, I have had a few experiences that have given me cause for serious reflection. The first was in the summer of 2017, when I found a parking ticket under my windscreen wiper after an evening out in the city of Bruges in Belgium. The message read 'No valid parking ticket'. It was my fault – I can't deny it. After I had given vent to my frustration by letting slip a blasphemous curse, a sympathetic Bruges resident asked whether I had been 'caught by the smart moped'. My jaw quite literally dropped. A 'smart moped'? Well, yes: apparently, there is a traffic warden who rides around Bruges on a scooter that can automatically identify parked cars without a valid parking ticket.

And it seems that this story is absolutely true. As part of its parking policy, Bruges is one of the first cities in the world to use new intelligent software to track down people who park without paying. This software recognises number plates and simultaneously checks whether the car in question belongs to a resident who holds a resident's parking permit. If not, the software instantly checks whether the holder of the number plate has paid for a parking ticket, either by SMS or using a different method. If they haven't, the miscreant can expect to be issued with an instant parking fine. This high-quality, advanced technology is giving rise to some confusion in the narrow medieval streets of Bruges. According to the company that supplies the cameras and software, traffic wardens can check only a few hundred cars per day, but these smart cameras can process no fewer than 10,000 vehicles over the same period. I must admit that I was rather impressed.

However, after the initial euphoria of my appreciation for this technology had died away, other questions came to mind. What about the traffic wardens? If a machine is ten times cheaper and more efficient than human labour, surely these workers must be afraid of losing their jobs?

And this is a question that has increasingly preoccupied me since this incident. In the spring of 2018 I was on a weekend trip to Monaco – after all, a tax specialist has to keep up to date with what is going on in the world – and I noticed that there were now hardly any check-in staff at Nice airport. They had been replaced by a very friendly machine that could explain to me in 15 different languages what I needed to do to weigh and label my luggage myself. Human labour had become a thing of the past. As a matter of fact, self-labelling turned out to be rather convenient and, what is more, they seemed to have got rid of the long queues I was expecting.

And you may also have noticed that cashiers in supermarkets are increasingly being replaced by self scanners, with a single employee supervising four or five of these automatic checkouts. Likewise in banks, where the traditional banker has long ago been replaced by your own laptop or smartphone, which you can use anywhere and at any time to make your monthly payments.

If it does indeed turn out that a substantial number of the jobs traditionally performed by human beings are now being done by robots and computers, what will become of the employees they replace? Are they being dismissed, or will they end up doing other jobs in the airport or supermarket? Given the large number of redundancies that have recently been announced in the banking sector, it is now clear what the impact of these technological changes has been on the staff in this industry at least. And, if the banking sector is a bellwether for other parts of the economy, isn't it likely that technological progress will give rise to social decline?

Anyone who examines the relevant academic literature will undoubtedly come across the paper on the future of employment in society by Carl Frey and Michael Osborne, both of whom are fellows at the University of Oxford. In their study in 2013, Frey and Osborne state that, over the next 20 years, 47 percent of current jobs will be at risk as a result of robotization and digitization. Other studies are somewhat less pessimistic. In March 2018, the OECD produced a report stating that 'only' 14 percent of the jobs in the OECD countries are at risk of being fully automated in the near future. However, it also stated that the way in which 32 percent of jobs will be done is set to undergo significant changes due to the use of technology in the labour process.

If more jobs are to be replaced by technology, there is a chance that the technological revolution will pose a danger to our welfare state. After all, social security is financed mainly by contributions relating to labour. And it will also give rise to problems within the tax system because approximately one-third of the federal government's total tax income is derived from income tax on wages. Job losses will therefore give rise to a loss of income for the welfare state. This brings us back to the main question of this chapter, namely that of whether we should impose tax on robots. If a taxi driver has to hand over part of his wages to the tax and social security authorities, but a self-driving car that does the same work makes no contributions whatsoever to the welfare state, then perhaps we need to reflect on how we can make robots a part of a fair tax system.

Danger to the welfare state

Although the results of research into the impact of the technological revolution on our jobs are extremely mixed, it's indisputable that there will be an impact and that jobs will be lost. And, of course, every job that does disappear is a personal tragedy, possibly made worse by the reason for the dismissal. To be let go due to poor performance or company reorganisation is one thing, but to lose your job because technology has completely taken over a certain role is another entirely, and the psychological impact of this should not be underestimated. But it is clear that the lurking danger lies not at the individual level alone but also at a macro level, and the government must take up arms against the negative consequences that the technological big bang will have for employment.

Because, if technology is going to cause certain jobs to disappear, this will also have an important impact on the financing and expenditure of the welfare state, possibly with disastrous consequences. After all, the current fiscal and parafiscal financing model of most OECD countries relies substantially on income from employment.

The importance of employment in the financing of the welfare state therefore also means that the welfare state may come under pressure as a result of the technological revolution. If jobs are lost, it goes without saying that the state will lose the fiscal and parafiscal income that goes with those jobs. Moreover, not only do job losses mean reduced income for social security, they also give rise to increased expenditure because the state has to provide the people who have been ousted from their jobs with replacement income in the form of jobseeker's allowance, at least in the short term. Therefore, it is clear that job losses as a result of technological changes will put our welfare model under pressure.

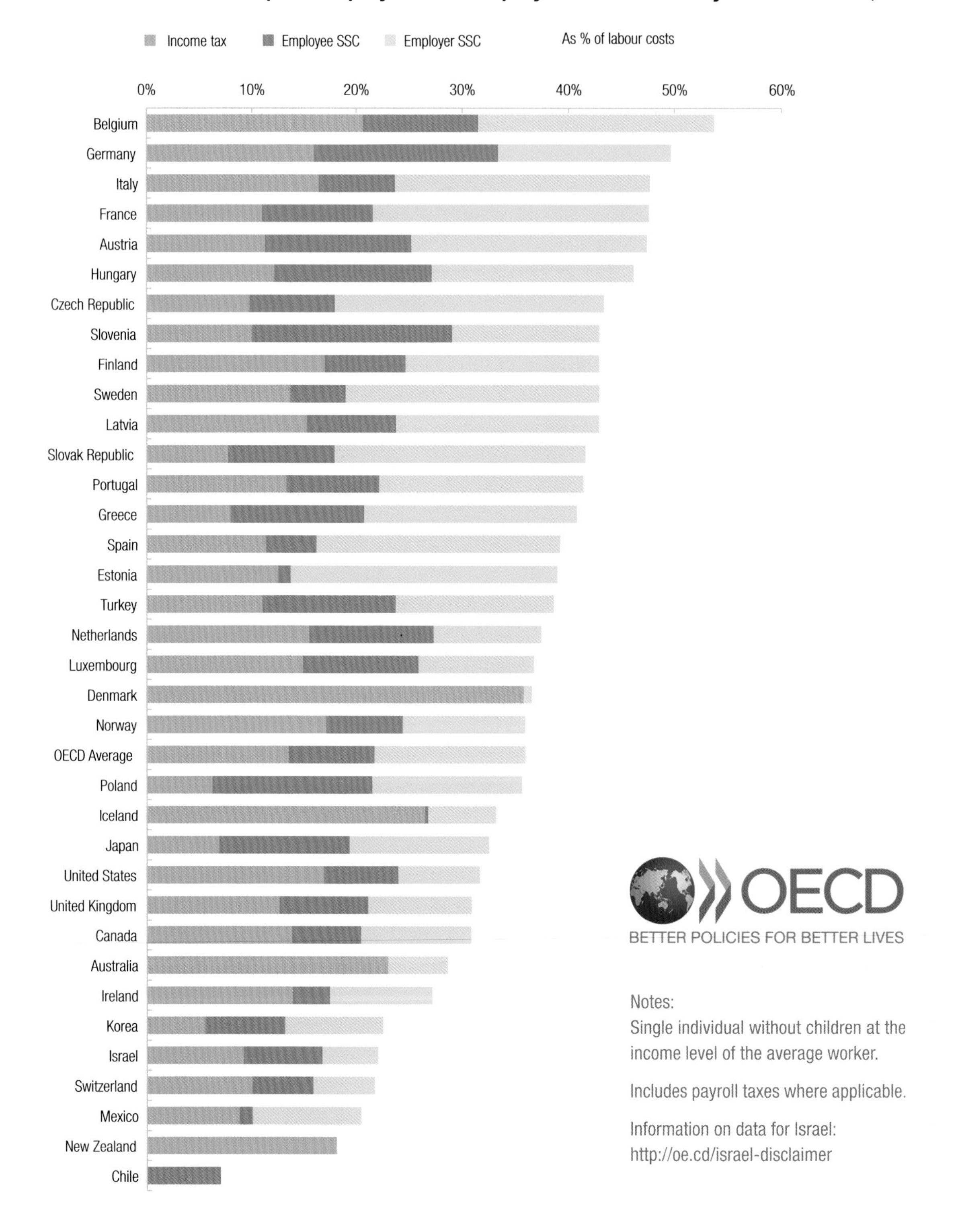

Notes:
Single individual without children at the income level of the average worker.

Includes payroll taxes where applicable.

Information on data for Israel: http://oe.cd/israel-disclaimer

On the other hand, intellectual integrity also requires a degree of nuance as far as the loss of tax revenue is concerned. An employer who decides to invest in employment-replacing technology because this is cheaper than labour will naturally make more profit, and that employer will also pay more income tax as a result. To return to my example from the introduction, if a parking company invests in number-plate recognition software because this is more efficient and cheaper than a traffic warden, that company will cut its costs and increase its profits, and will have to pay corporation tax on the additional profit it makes. But, since we know that the fiscal and parafiscal burden on employment is much higher than the burden on operating profit, it is clear that the additional tax on the operating profit can never compensate for the loss due to redundancies.

The conclusion is therefore abundantly clear. If the technological revolution causes jobs to disappear, this will adversely affect the resources of the welfare state. If the government wants to keep the welfare state up to standard, something will need to be done about the fiscal and parafiscal systems to ensure that the welfare generated by robots will benefit society as a whole. That is only logical, because why should a waiter, cashier or driver hand over part of their wages to the tax and social security authorities, but a robot doing the same work pay no tax at all?

How does one tax robots?

Ensuring that the prosperity created by robots helps to finance the welfare state is an attractive idea. But how do you tax robots? This is far from self-evident because there are various factors that need to be taken into account. For example, real industrial robots have a high visibility and can be taxed relatively easily based upon their power, but what do you do, for example, with computing power or algorithms, which are much less tangible? So, finding a basis for taxation that is generally acceptable and applicable is not as trivial a task as it seems.

In the spring of 2017, Microsoft's founder, Bill Gates, floated the idea of imposing a tax to compensate for the lost income tax whenever an employee is replaced by a robot. The proceeds of such a robot tax could be used to finance training courses to educate human manpower to help people adjust to technological change. This would allow factory workers, drivers and cashiers who are replaced by robots to establish new careers as health workers or teachers, or work in other non-robotised jobs.

At first glance, this seems like an attractive idea, but, when examined more closely, this form of taxation is not practicable for legal reasons. In my example of the parking company that replaces employees with number-plate recognition software, this would result in the parking company having to pay an additional robot tax equal to the personal income tax that the dismissed employees owed on their pay. This system may be practicable if a company actually dismisses employees and replaces them with technology. But what should the government do when a new company is founded that uses technology right from the outset and therefore never had any employees to dismiss? In this case, the robot tax could not be imposed, which, of course, would give rise to fiscal discrimination between old and new companies.

And what about a flat-rate tax on robots and computers: would that be viable? Well, to a certain extent, we already have such a tax, albeit at local level. A considerable number of municipal councils impose a tax on motive power – i.e. a tax on motors calculated per kilowatt – or on ATMs. Taxing services is also very popular with municipal councils, enabling them to tax the provision of tourist accommodation through internet platforms such as Airbnb or taxi services through companies such as Uber. But, again, this form of taxation is not really efficient. Dividing technology into categories and subjecting it to various types of flat-rate taxes gives rise to fiscal chaos and administrative red tape. What's more, all those taxes need to be checked, which puts a heavy burden on the taxation authorities. So, that's not an attractive alternative either.

So what would an efficient robot tax look like? Perhaps we are overthinking things and just need to turn the argument around. The principle underlying the proposals to implement a robot tax is that companies using technology instead of employees should be fiscally penalised for doing so, paying tax as a form of compensation. So, what if we were to instead fiscally reward companies employing human beings? Isn't that worth considering? Under this system, all companies would be taxed equally based upon the same corporation tax rules, but companies that employ people would be granted an additional tax benefit on payroll costs over and above the normal deduction as a business expense. In our current fiscal model, companies that invest in technology and other assets are already allocated additional tax benefits by way of an *investment deduction or innovation deduction*, which is added to the fiscal write-off. This allows companies to recover more than 100 percent of the cost price of the amount they have invested. In a sense, this is absurd because such investments can lead to job losses and that means we, as a society, are actually incentivising these redundancies. So, it is clear that the tax system is shooting itself in the foot. We can turn that system around by providing additional fiscal rewards only to those companies that employ people, either in the form of an *employment deduction*, in which more than 100 percent of salary costs become deductible, or by a direct *job deduction* under corporation tax.

Under this system, companies with a low employment rate would bear a greater tax burden than companies with a higher one. The difference in tax burden based upon the employment rate can be justified on the grounds that companies with a high employment rate contribute more to financing the welfare state by withholding tax and social security contributions from employees' salaries than companies with a low employment rate. The above model enables both domestic companies and foreign companies with a permanent base in the country (which are therefore subject to the country's corporation tax), to contribute to the financing of the welfare state through a diversified system of taxation, either directly through corporation tax or indirectly through personal income tax by fiscally supported job creation.

However, foreign companies that are not permanently based on national territory are not subject to the country's corporation tax and therefore contribute nothing at all to the financing of the welfare state through taxation. For example, this is the case for tech companies such as Facebook, Google and Amazon, which only have '*users*' or '*customers*' on national territory. They generate profit through the users and customers in a country, but do not pay any income tax. Under a fair taxation system, it is self-evident that these companies should contribute to the prosperity from which they benefit. These companies should also be subject to a specific robot tax, in the form of a digitax. At the start of 2018, the European Commission launched a proposal to start taxing such companies on the profits that they make in every EU Member State from sales, subscriptions and the commercial use of personal data. This is a very valid proposal in the context of a fair and forward-looking fiscal system.

Conclusion

In view of the inevitable robotization of the economy, the government will be compelled to consider what the fiscal system of the future will look like. We need to move away from the idea of income from work as the central pillar supporting the fiscal and parafiscal system. If jobs have to make way for technology, the government will need to shift the focus of taxation from the income on employment to profits from human or technological work. In a sense, we need to return to ancient Rome where the Romans taxed slaves rather than their work. Today, we need to ensure that our modern slaves – robot is actually the Czech word for a slave – make their contribution to the financing of the welfare state. Once the internet of things has become a reality we will inevitably also need to move to the taxation of things.

Amazon has more than 100,000 robots driving around their warehouses, bringing racks to order pickers, who then take out the products and put them in boxes.

©Amazon Robotics

SUSTAINABILITY

The Opsoro platform is an open source platform for social robots developed by Ghent University.

©UGent-Elig

MUMMY, CAN I BUILD A ROBOT?

By Joachim Mathieu and Jef Van Laer

The Vrije Universiteit Brussel (VUB) has been a pioneer in the field of science communication in Flanders for more than thirty years, and in 2008 a collaboration between the robotics department and the scientific communications expertise centre gave rise to RoboCup Junior. Over the past ten years, this robotics competition for young people has rapidly developed into one of the most successful outreach initiatives with a STEM (Science, Technology, Engineering, Mathematics) focus in Flanders, with involvement of almost all Flemish higher education institutions.

Attracting young people into STEM

Flanders, like many other industrialised countries, suffers from a major shortage of technicians and graduates in the exact sciences. Employers and employers' organisations have for years been sounding the alarm and calling for these shortages in the labour market to be addressed. At the beginning of the 1990s, this problem found its way onto the agenda of European policy discussions, resulting in the Lisbon objectives in which EU Member States committed themselves to becoming part of the most competitive and dynamic knowledge-based economy in the world.

Such an ambitious plan required effort on the part of the Member States, and Flanders had to significantly increase its investment in research and innovation, and work harder to attract young people into STEM jobs and careers. It soon became clear that this would not be an easy task, and the issue is still highly relevant today. For example, technical roles make up half of the top ten bottleneck professions on the list published each year by the Flemish Employment and Vocational Training Agency (VDAB). Although there has been a slight increase in young people opting for STEM routes in recent years, the numbers are still far too low, and there is a particular shortage of girls. As a result, we are still seeing too few STEM graduates leaving our higher education.

Research shows that, like adults, young people are positive about science and technology, but they are much less enthusiastic when it comes to science education, and this lack of interest in STEM subjects has an immediate knock-on effect upon their choice of further study and careers. Even if they obtain good marks in these subjects at secondary school, they rarely choose a STEM field of study in higher education. The age at which young people come into contact with STEM seems to be a major factor, with interest first developing at an early age and being established by the age of around fourteen in most youngsters. So we need to work out how to engage young people in science and technology at an early age.

Over the past five years in Flanders, STEM seems to have become the acronym that is on everyone's lips. Not only do schools use a STEM label to evaluate their performance, there are also hundreds of STEM academies, where children and young people can do STEM activities in their spare time. In August 2018, the Flemish Minister for Work, Economy, Innovation and Sport, Philippe Muyters (N-VA, New Flemish Alliance), announced that he would be making extra support available so that every municipality in Flanders could have its own STEM academy.

STEM activities need to go beyond technical crafts, with a good STEM activity combining aspects of each of the four pillars (science, technology, engineering and mathematics) and giving young people space to work independently using inquiry-based learning. This integration of different scientific and technical disciplines is particularly evident in robotics.

Learning with robots

The increasing presence of robotics in our daily lives and the fact that there are so many robotics kits available for educational purposes makes this field ideal for teaching science and technical literacy to young people. It is also unique in the way it unites the different components of STEM, and has the added advantage of being intriguing to young people and adults alike.

Robotics brings together various scientific fields, providing an opportunity to apply the principles of algebra, trigonometry, materials science and design, as well as the more obvious mechanics and programming.

But it's not enough to simply take robotics as a general theme. You won't get the results you want by simply giving a group of youngsters a robot construction kit like Lego Mindstorms and leaving them to it, seeing this approach is more likely to turn young people off STEM than to get them hooked. A good framework, effective guidance and a clear goal are crucial to making robotics a useful tool in STEM education.

The VUB has been using robotics as part of STEM education for more than ten years (in fact, since before the acronym was even in use in Flanders) and in 2008 its Mechanics department joined forces with the Science Outreach Office of the VUB to set up a robotics project for young people aged 8 to 18. A robotics competition seemed like the ideal way to awaken enthusiasm for STEM amongst this broad target group. Whilst definitely not the only way to pique the interest of young people – workshops, science camps and visits to companies where robots are used have also yielded good results – competitions and the opportunities for project-based working they provide mean that participants have to keep a clear goal in mind and can work hard whilst still having fun. Integrated STEM projects also have the important advantage that they attract girls to scientific and technological subjects, with boys and girls alike finding robots interesting and intriguing. So, robotic competitions are in no way *boys' clubs*.

Children watch enthralled during a visit to the robotics lab of VUB-Brubotics.

©Jeugd, Cultuur & Wetenschap

A ROBOT FOOTBALL TEAM

RoboCup is an annual competition between companies and research institutions set up with the aim of sharing knowledge and facilitating research. Its ultimate ambition is to develop a team of football-playing robots that can compete against the winners of the human football World Cup by 2050.

What kind of robotics competition?

So, how do you go about organising a robotics competition? The first job was to choose a concept. Since a number of these competitions already existed in other countries, it was not necessary to come up with a totally new concept from scratch. Rather, the next question was which concept to choose from all those that were available. For a number of reasons, we finally opted to hold a Flemish version of RoboCup Junior. This is an international competition for children and young people and the little brother of RoboCup. It has been around since 1998.

The main reason for choosing RoboCup Junior was that it is not tied to a particular type of hardware or software, leaving participating teams free to choose which robot package to work with or even to design and add their own parts. This encourages creativity and brings greater diversity to the solutions developed by the teams in response to the challenges. Participants can also develop their project over a number of years by starting with an easily accessible platform such as Lego Mindstorms and then switching to more complex platforms that offer a greater challenge later on. In other words, RoboCup Junior allows the difficulty level to be adjusted at will.

Although diversity is to be encouraged in the robotic creations, the competition itself needs to have a certain degree of continuity. Allowing the number of disciplines to proliferate or completely changing the assignments or rules every year would make things unnecessarily complicated for schools and increase the costs for the participating teams. RoboCup Junior therefore strives for consistency in terms of disciplines and rules from one year to the next.

The international context was another factor in our choice of RoboCup Junior, as it gave us a framework for the exchange of ideas and best practices and allowed us to participate in regional and world championships.

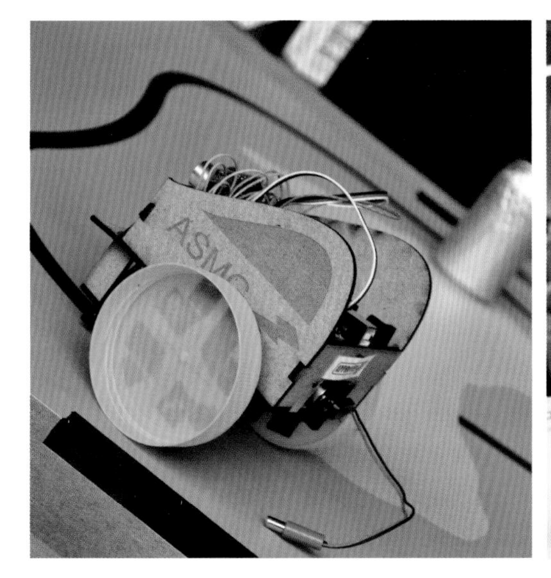

RoboCup Junior is open to all robotics packages and also welcomes creations built from scratch.
© RoboCup Junior

The assignments themselves were another plus point as they encourage the use of sensors, without which the robot would be no more than a machine.

Finally, we wanted our competition to have a positive message. Rather than programming robots to fight each other, we wanted participants to design robots that can make a positive contribution to society and are thus in line with the research goals at VUB. The RoboCup Junior assignments are not violent, with the three international disciplines being *Onstage, Rescue* and *Soccer.* There are also subdisciplines within these categories.

One hundred teams

The first event was organised in the spring of 2009 with the support of a consortium of partners from higher education and industry throughout Flanders. Working in this way helped to encourage teams from all over the area to participate. Support for all teams in the form of contact and workshops further boosted the geographical distribution of participants. Links with industry were also nurtured, with companies such as Honda, Asimo Studios, iRobot and National Instruments backing the initiative. And RoboCup sought and found a third type of partner in the press, with TV channel Ketnet piquing the interest of its younger audience by reporting on the competition regularly and featuring it extensively in its children's news programme Karrewiet. EOS magazine also kept its readers informed about the initiative.

Young people aged 8 to 18 years were able to register teams for the *Onstage* and *Rescue* disciplines, and didactic support was provided to make sure the teams had sufficient guidance and expertise available to them. This support took the form of regional *teach the teacher workshops*, a teaching package, a handy robotics book and manuals on both the competition and the use of Lego's NXT package.

After that first event, interest grew quickly among schools and teachers and capacity has now been significantly expanded. Ten years on, around a hundred teams can participate in the competition every year, with these places typically being filled within a few weeks of registration opening.

Many schools organise their own internal championships in the run-up to the event, with the best teams representing them in the actual competition. Like the First Lego League and the Robot competition, RoboCup Junior is extremely popular, attracting teams from primary and secondary schools alike. It can also count on the interest of teams participating as a group of friends or supported by extracurricular activities such as Coderdojo, making it an important resource for informal learning.

Performing or rescuing

Teams participating in RoboCup Junior can choose between three disciplines. One of these is *Onstage*, in which teams develop a complete act, with one or more robots providing the central spectacle. Creativity and interaction between man and robot are key aspects of this assignment. Although *Onstage* tends to be regarded as an entry-level discipline, programming, sensor use and problem-solving are strongly encouraged.

The *Rescue* discipline has two subdisciplines: *Rescue (beginners)* and *Rescue (advanced)*. The goal of both of these is for the robot to autonomously follow a route, overcoming obstacles on the way, and then 'rescue' an object from a demarcated zone. In contrast to the beginners' variant, *Rescue (advanced)* takes place on a course made up of an ever-changing combination of tiles. The various obstacles, such as speed bumps, complicated junctions and objects in the middle of the course that need to be avoided, are also more challenging than those in the *Rescue (beginners)* category. However, even the basic variant is a serious challenge and the teams need to work their way towards possible solutions gradually using trial-and-error and teamwork.

The winning teams can participate in the world championship, which is organised annually in collaboration with RoboCup. In the past, Flemish teams have taken part in the disciplines *OnStage* and *Rescue* in Graz, Eindhoven, Istanbul and Leipzig.

RoboCup Junior is all about experiential learning based on a challenge designed to inspire participants. Robotics competitions can therefore teach participants about more than just technology and science. The project allows cross-curricular work, boosts technical self-confidence and encourages participants to exchange knowledge, communicate and engage in teamwork. They also learn that there are several possible ways to approach a challenge and that they can inch their way closer to a solution by a process of trial-and-error.

The popularity of robotics competitions such as RoboCup Junior and First Lego League bears witness to the appeal and usefulness of this type of STEM education. It also demonstrates that there is a demand for project-based support in both primary and secondary education.

More information about the contest: www.robocupjunior.be

More information about the Science Outreach Office, affiliated to the Erasmus University College Brussels (EhB) and VUB: www.wtnschp.be/sciencesays

Rector Caroline Pauwels drinks a fresh smoothie made by a smoothie machine from Alberts, a company founded by two former engineering students of the VUB.

SO, WHERE ARE THOSE ENTREPRENEURS IN ROBOTICS?

By Prof. Dr. Thomas Crispeels, Lennert Vierendeels
and Prof. Marc Goldchstein

New markets and technologies are the ultimate playground for entrepreneurs, and we are seeing a whole series of 'new' companies springing up, founded by *robopreneurs*: American companies like iRobot, Boston Dynamics, Ekso Bionics and Rethink Robotics, Japanese ones like Cyberdyne, and European ones like Universal Robots, Aldebaran Robotics and Blue Frog Robotics. Moreover, the first robotics companies are already hard at it in Belgium. Zorabots from Ostend, for example, has become a world leader in the field of social robots. But despite promising early success stories, the big boom in Flemish robot companies has yet to materialise.

Do we even need those robopreneurs?

The combination of technological innovation and entrepreneurs can have far-reaching consequences, especially in the case of a technology with a huge impact on productivity. We only need think of all the changes to our society over the past century and a half because entrepreneurs picked up on the combustion engine, the computer chip and the internet. These innovations affect not only the economy, but every aspect of society, such as how we interact with each other, our values and ideas, and the way we engage in politics.

The first type of economic impact an innovation can have relates to its creation. Innovations give rise to new opportunities which, thanks to entrepreneurs, lead to new initiatives, new services, new products and even new industries. Take the development of the digital camera, for example. By making the film roll redundant and facilitating the move to digitised images, it not only led to the development of new and better cameras, but also spawned numerous new companies and initiatives in the field of digital photography: online print shops, software packages, portable storage media, social media.

In other words, a revolution like that of robotics allows a whole range of new possibilities to emerge and be tried out. For example, we can already use robots to vacuum our house, make a smoothie or film our mountain bike trip from a distance.

But this creation also has a downside: destruction. New technologies, products and industries often make existing companies, products and jobs redundant. Whereas, in the past, a single bank transfer called for a large amount of human labour on the part of banking staff, we can now do it ourselves in seconds via an app. The contraction of the branch network of most banks (destruction) is thus a logical consequence of the digitisation of our society (creation).

Robotics entrepreneur Pierre Cherelle founded the company Axiles Bionics for a new generation of bionic feet, developed by VUB-Brubotics.

This cycle of 'creative destruction', as Joseph Schumpeter called it in 1942, is one of the core elements of our current capitalist economic system and ensures the indispensable rejuvenation of our economic fabric. Our entrepreneurs and their start-ups are an essential part of this rejuvenation.

Who will become a robot entrepreneur?

The story begins with an entrepreneur who has an idea. And it's a myth that the best ideas arrive out of the blue: stories like that are often invented after the event for or by the media. For example, it is all too easy to forget that Ebay's founder, Pierre Omidyar, worked for digital and e-commerce companies for seven years before launching Ebay. Bill Gates, too, had been addicted to computers and programming since the age of thirteen. He founded Microsoft in 1975, when he had reached the grand old age of twenty.

The entrepreneurs behind some of the new robotics companies had also been working with robots for years before they made the leap into entrepreneurship. Rodney Brooks, one of the best-known robopreneurs, had been professor of robotics at MIT – and guest researcher at the

The social robots Billy and Cruzr from Zora Robotics.
©Zora Robotics

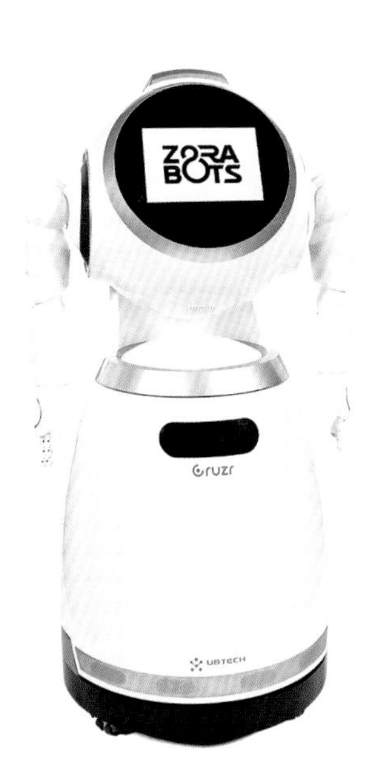

VUB AI Lab – for many years before breaking into two new markets with iRobot (vacuum cleaner robots) and later Rethink Robotics (industrial collaborative robots). The founders of Zorabots in Ostend had also been fascinated by robots, and had followed the sector closely, for years. Bruno Maisonnier, the founder of pioneering social robotics company Aldebaran robotics, combined his experience in the business world with a lifelong passion for robotics.

In the field of high-tech entrepreneurship, we often make a distinction between two main categories of entrepreneurs: product starters and technology starters. Product starters are often professionals who have many years of experience in the industry behind them and make a connection between a new, emerging technology and an existing customer need. For example, cameramen who set up their own companies specialising in filming sports events and commercials using drones.

Technology starters, on the other hand, often take a new technology as their point of departure. For example, every day hundreds of researchers at Flemish universities are working on the very latest robotic technologies. They develop a component or concept and then look for a promising application. A good example is the VUB spin-off currently being set up, Axiles Bionics, where fundamental research into more efficient actuators – which can be thought of as the muscles of robots – led to these components being used in exoskeletons and bionic prostheses. It is then down to the researchers, who often have little affinity with the industry in question, to go into business and to further develop and market the technology.

The university as a knowledge centre

Suppose the entrepreneur has identified an opportunity. What other ingredients are needed before he or she can get down to business?

The entrepreneur's story often starts with knowledge: knowledge about a certain opportunity or technology, and knowledge about the customer, market, etc. This can be acquired through experience and training or is sometimes found in written sources such as scientific articles, manuals or patents. It is no coincidence that many new robotics companies have strong links to universities. It takes a lot of expertise to design, produce and market a robot: engineers are needed to make the physical robot, artificial intelligence specialists to make the robot smart, psychologists or sociologists to get the robot to work with people, and entrepreneurs to put the robot on the market. Universities are one of the few places to bring all this together under one roof.

An effective and well-balanced team

As soon as entrepreneurs have an idea, they start looking for the right people so they can to put a team together – it's an illusion to think that a single individual can do everything. Our ideas about this are often distorted because in the United States the focus is usually on the hero of the story, for example Steve Jobs, Marc Zuckerberg, Bill Gates or Elon Musk. But make no mistake, all those top entrepreneurs were assisted by an excellent and well-balanced team right from the outset. Flemish research has shown that high-tech entrepreneurial teams with complementary knowledge have a more regular growth path than teams with less diverse backgrounds. Finding all these people is often one of the most difficult tasks for young entrepreneurs: in Flanders, for example, the supply of experts in the field of robotics and artificial intelligence is limited.

Clusters, a place to network

Another way to find knowledge is by networking. It's how good entrepreneurs mobilise their network, gain faster access to knowledge and reach customers or financiers. This also explains why many high-tech industries 'cluster': for example, in the US, most robotics companies are located in Detroit, close to the automotive industry, or in Boston, near the world-renowned universities like MIT and Harvard. In the Netherlands, most of these companies are located in 'Robovalley', near Delft University of Technology and, in France, Paris is the hotspot thanks to the many incubators affiliated with the Paris universities. In Flanders we have a similar cluster of biotechnology companies, but we still lack a robotics cluster.

Finding start-up capital

That brings us to one final important resource: money. In a high-tech sector such as robotics, in particular, entrepreneurs can seldom put up the money they need to transform an often immature technology into a market-ready product themselves. After all, a robot is a combination of hardware and software, both of which require a research and development path. The start-up capital needed can soon add up to several hundred thousand euros.

Robotics entrepreneurs who aren't in a position to sell their products straight away, must therefore look for external investors. Banks won't usually lend money to risky companies, but fortunately there are specialised private investors who are willing to invest in *high-risk/high-potential companies*. In the US, this venture capital is often more readily available. Ekso Bionics, which builds exoskeletons, and BionX, which develops robot prostheses, have both raised around

70 million dollars. Rethink robotics, one of the pioneers of the fast-growing cobot market, raised more than 150 million dollars, but recently had to file for bankruptcy due to a lack of financial resources.

What can the government do?

Every government wants to welcome the new Silicon Valley into its country. A vibrant tech scene creates valuable quality jobs and extra tax revenue. How can governments encourage the creation of new industries in their territory? They can use regulations and incentives to create an ecosystem that supports high-tech entrepreneurs and thus increases the chance that *the next big thing* will happen on their home turf. But they can also prevent entrepreneurs from being scared off. It is clear that the development of a favourable climate calls for a long-term strategy, a clear vision and the necessary (political) courage.

Betting on knowledge

Since knowledge is the starting point for everything, most policy plans revolve around building and supporting a knowledge economy. The government has a crucial role to play in terms of the funding of research programmes. In the United States, for example, funding from the Defense Advanced Research Projects Agency formed the basis for one of the world's best-known robotics companies, Boston Dynamics. Closer to home, the European Union is investing heavily in robotics research, which has already given rise to a string of robotics start-ups. Moreover, Belgium has a very favourable tax regime for knowledge-creating companies and provides subsidies for research and development within universities and companies. Its government has also ensured that all results and inventions developed in universities are the unquestionable property of those universities. Universities are therefore responsible for the commercialisation of their research results.

Governments also need to put in place a system of stable, clear and enforceable intellectual property rights. This allows entrepreneurs to be confident that their products can be launched onto the market and that their patents will be recognised in court.

Finally, the government can also bring its influence to bear through the legislative framework. Entrepreneurs like to avoid risk and want to be sure that their product or service can actually be launched onto the market. For example, the Belgian parliament has allowed (limited) tests with self-driving cars.

More of the right people are needed

The government can also help by ensuring that enough highly skilled workers are available. This can be done by providing high-quality training to its own population and by facilitating international mobility: Flanders and Brussels are well placed to attract foreign talent. Although Flemish universities offer excellent courses, demand for experts exceeds supply, so if we want to develop a robotic ecosystem, we need to focus more on encouraging people to train or retrain in robotics.

Networks are a personal thing and, as stated earlier, represent a powerful and relatively cheap instrument for entrepreneurs. The government can aid networking by supporting cluster federations and network organisations, and by encouraging or even requiring active cooperation between organisations in subsidy channels.

A question of capital and risk

The government can support start-ups by providing subsidies or subordinated loans. But many governments are all too aware that it is much more efficient to leave the selection of projects to the private sector. Professional investors evaluate projects on a daily basis and only finance the best. Sometimes a government can lower the barriers to these investors by setting up its own investment company (such as PMV in Flanders) or by co-investing in private investments, which may focus on a particular sector or technology, as is the case for the Arkimedes fund and Biotech Fund Flanders. The major European cobot pioneer Universal Robots was founded in Denmark with the support of the Danish government.

The entrepreneurs are there, but you can't see them

Now we come to our key question. Where are all these robotics entrepreneurs? Well, they're there, but you don't often get to see them. Many technological innovations are almost invisible to us as consumers. When Henry Ford decided to change how the Ford Model T was produced, he didn't change anything about the concept or technology of the car. He simply automated the production process. Today, production lines are made up of robots of all shapes and sizes, supplied by various companies and incorporating components and technologies from thousands of manufacturers from all over the world.

Since a lot of innovations are invisible, this also means that a great many opportunities exist that are invisible to us. The first robotics companies have now been identified in Flanders. If the robotics industry is to flourish and retain a permanent presence in our region, an integrated and focused policy is necessary – at a time when the added value of the sector (in terms of jobs, tax revenues, etc.) is barely measurable. This calls for vision and courage.

DESTROYED BY ITS OWN INVENTION

The most noteworthy (and most commonly cited) example of 'creative destruction' can be found in the domain of digital photography. The first true digital camera was invented in 1975 in the labs of Eastman Kodak, the company that used to be synonymous with the camera. However, the company proved unable, or perhaps even unwilling, to exploit that invention. After all, the quality of its digital photos was very low and the company was doing very nicely in the market for analogue photography. Just over thirty years later, in 2012, the company went bankrupt. It's a story of how the product of a company's lab led to the destruction of its own parent company.

ROBOTS: FRIEND OR FOE OF THE ENERGY-EFFICIENT SOCIETY?

By Dr. Ir. Tom Verstraten

These days, when we talk about sustainability our first thought tends to be about energy consumption. Climate change is increasingly dominating the news and we still get most of our energy from 'dirty' sources. Further developments in green energy could be part of the solution to this problem, but the simplest way to save the environment is still the best: reduce our energy consumption. Robots can't do anything without energy, so we need to ask ourselves whether more robots mean higher energy consumption overall.

They are coming!

Today's robots are mainly found in factories, and Belgium is a forerunner in this field. But robots represent only a very small proportion of the total energy consumption of a factory. For example, car factories are one place where you are likely to find lots of robots at work but even here they account for barely 8 percent of total energy consumption – and the figure is only around 1 percent for the industry as a whole. So, as yet, it's not such a big deal.

Of course, these figures will rise as the number of robots in society goes up, and we are seeing an increase in robot sales of 15 to 25 percent per year. Whereas robots used to be almost the exclusive preserve of factories, they are gradually finding their way into our homes, with many Belgian families already being the proud owners of robotic vacuum cleaners or lawnmowers.

And, what about robotic butlers to bring us a glass of water or put out the bins? Anyone following the latest developments in robotics knows that this idea is not as far-fetched as you might think. In other words, we are in the early stages of a true robotic revolution. So now seems like a good time to give some thought to the subject of energy-efficient robots.

Energy consumption and energy efficiency

Energy efficiency is a term used to describe the relationship between the amount of energy we put into a machine and the amount of useful work we get out of it. Engineers express this ratio in terms of efficiency: the higher the efficiency, the more economical the unit will be to run.

Robots are usually powered by multiple electric motors, which are generally very energy efficient, with figures of 90 percent and above being perfectly normal. Yet the efficiency of the robots themselves is often much lower. In many cases only a fraction of the electrical energy supplied is converted into useful work and efficiencies usually lie around 10 percent. This means that the amount of energy that is lost is nine times greater than the amount actually used to drive the robot.

So what is it that makes robots so inefficient? To answer that question, we need to look at how motors are used in robots. Motors achieve their maximum efficiency when they run at a specific speed and deliver a specific force. Robots often have to carry a heavy load while almost stationary or, conversely, perform very fast movements while they are barely loaded at all, meaning that these optimal conditions are hardly ever in place. And these are just two ways in which motors are used inefficiently. So, it should come as no surprise that, considered as a whole, robots achieve particularly poor efficiencies, which is bad for their owners' electricity bills and, by extension, bad for the environment.

But that's not the only concern. Many robots – rescue robots, robotic butlers, active prostheses and exoskeletons to name but a few – can't simply be plugged into a power socket and need batteries to keep them moving. In these applications, high energy consumption translates into the robots having to lug large batteries – that also need frequent recharging – around with them. This makes them heavy and cumbersome, which in turn decreases the smoothness and safety of their movements, and causes their power consumption to rise still further. Reducing energy consumption is therefore a high priority for such robots.

A MATTER OF EFFICIENCY

The efficiency of a unit is the ratio of the amount of energy it delivers to the energy it consumes. Energy is expressed in joules and it takes around 2,000 joules of energy to lift a mass of 100 kilograms to a height of two metres. If you use an electric crane, it takes slightly more electrical energy to lift the load because you have to provide enough energy to compensate for any losses as well as to actually do the work. So, if the crane consumes 2,500 joules, the efficiency of the crane is 2,000/2,500=80 percent. In other words: 80 percent of the energy you supply is put to good use and the rest is lost in unusable heat.

The energy efficiency of a spring

The simplest way to reduce the power consumption of robots is to make them more efficient. Another booming industry – the ICT sector – has proved that such an approach can pay off. ICT now accounts for almost 10 percent of Belgium's demand for electricity and you would expect that share to be on the rise. But a Dutch study has indicated that the ICT sector's energy consumption remained stable, and even fell slightly, between 2007 and 2013. There are several reasons for this, but the increased efficiency of the devices in question is one of the most important factors.

Researchers around the world are therefore experimenting with new concepts to make robots work more efficiently, using approaches such as optimising their movements, having robots work together from a shared power supply, making them lighter or using innovative designs.

One interesting solution currently being investigated is based upon the use of springs and other elastic elements. These are fairly easy to tension, and release a large amount of energy when this tension is relaxed. Just think of archery: fundamentally, a bow is a rope stretched between the two ends of a flexible stick. This simple device can launch an arrow at great speed, allowing it to cover distances of hundreds of meters. If you tried to throw the same arrow with your hand alone, you are unlikely to get it further than perhaps ten metres. Another example is the pogo stick. By compressing the stick's spring at the right time you can achieve impressive leaps: the current world record is 3,378 meters!

Elastic elements are also an important reason why people can move so efficiently. Thanks to their elasticity, our muscles and tendons are highly effective at transferring energy to the environment. This enables us to perform explosive movements such as jumping or sprinting, and to walk very efficiently. Springs are also used in many walking robots, sometimes with astonishing results: there are robots that can descend a slight slope naturally without a motor or batteries. In prosthetics and exoskeletons, they can significantly reduce power consumption. These are just a few examples of the potential for the use of springs in robotics.

Robots can help to save energy

So far, we have only considered how to reduce the energy consumption of robots, but our ambitions shouldn't stop there. Although robots themselves consume a certain amount of power, they can also help us to use energy more efficiently. How? In the first place, robotics and artificial intelligence can help us to cut our usage.

Imagine a factory hall populated solely by robots. Lighting? Not needed, because if robots are performing the same movement time and time again, they don't need to see. Heating or air conditioning? Robots don't need those either. And that's not all, because there may also be savings to be made in the work itself. For example, a welding or painting robot often consumes less energy than a human being doing the same job. This means lower energy bills for the factory owner, and good news for the environment.

There are also opportunities for savings in the field of mobility. In some cases, it may be possible for work that is currently done manually, to be taken over by robots, with the employee's job limited to controlling the robot. This can be done perfectly well from a distance – or even from home – and will mean that a large part of the working population no longer has to travel to work, reducing the need for commuting and the associated pollution and congestion. This will not only give rise to considerable energy savings, it will also be beneficial to society as a whole.

Still want to drive? In future you may find yourself using a self-driving car. These can be programmed to make driving as energy efficient as possible by the sparing use of the accelerator and brakes.

Self-driving cars may also allow traffic flow to be regulated more efficiently. For example, a self-driving car can drop you off at your destination and then take itself off to the nearest car park, making the tiresome process of driving around trying to find a parking space near the door a thing of the past.

And if we look beyond the field of robotics, even more is possible. Smart sensors and computers can help us fight energy waste, for example by continuously monitoring the temperature and the number of people in a room to optimise the settings of our central heating. This could prevent the heating from being turned up too high or from unoccupied rooms being heated.

Innovative solutions can also save energy on the streets. In Norway, for example, street lights were installed that automatically dim when there are no cars around. A similar idea could also be applied to waste collection by fitting bins with sensors and connecting them to the Internet. This would make it possible to determine exactly which bins need to be emptied when, dispensing with the need for the garbage truck to drive past every bin, and making smelly, chock-full bins a thing of the past. If we also use self-driving garbage trucks, a truly efficient waste collection system could be created.

Smart working and smart charging

Not only do robotics, AI and automation offer opportunities for energy savings, robots can also play a crucial role in environmentally friendly energy generation. The big problem with so-called 'green' energy sources is that you can't control when they supply energy: solar panels only produce electricity when the sun is shining and wind turbines are useless on a still day. This means there is often a mismatch between the electricity generated and our energy needs.

The solution seems obvious: store the excess energy for use in times of high demand. But that's not as easy as you might think. The most effective – and most commonly used – method is pumped storage. When excess electricity is available it is used to pump water to an upper reservoir; this water is then released when demand is high, allowing it to flow through a turbine and generate electricity. Pumped-storage plants allow large amounts of energy to be stored at a low cost, but take up a lot of space and represent a major investment.

So, are batteries the answer? Although it is indeed possible to store excess energy from the grid in this way, very large and expensive battery banks are needed. But scientists think that the batteries in our electric cars may represent the solution to our future energy storage needs. If everyone were to switch to an electric car, this would give rise to a huge amount of battery capacity spread across many places. Smart charging systems would make it possible for us to charge our cars at times when excess electricity is being generated. That is often at night, which is precisely the time when most cars are plugged in.

Robots can also be part of this move towards smarter energy consumption. Domestic robots – such as the robotic vacuum cleaners and lawnmowers we are already familiar with, as well as the robotic butlers of the future – work on batteries. So, like cars, they can be charged when electricity production peaks. In other words, when you plug in your robotic vacuum cleaner it waits for a signal from your electricity supplier before it starts charging.

Robots in factories can be connected to the electricity grid, so there are no batteries to charge and discharge. But these robots, too, can help smooth out the demand for electricity. When demand falls below supply, the robots can be brought on line and production started up. When it increases, the production line can be halted. This process can be fully automated and has the added advantage that robots are never going to complain about irregular working hours.

Seize the opportunities!

Robots are almost certain to make up an increasing proportion of our energy consumption in the future, but this need not increase our usage overall. Innovative solutions will lead to robots consuming less and less power, and robotics, AI and automation can be used to make our society more energy efficient. This technology is currently in development, with much already on the market. So it's up to our policy makers to pick up the ball and run with it.

WILL THERE BE A PURPLE BAG FOR ROBOTIC WASTE IN THE FUTURE?

By Dr. Ir. Joost Brancart, Prof Dr. Ir. Guy Van Assche and
Prof. Dr. Ir. Hubert Rahier

What effect do automation and robotics have on our waste production? What impact will the robotic revolution have on the sustainability of our society? It goes without saying that the production of robots calls for a huge amount of materials. Increasing automation and rising robot numbers will lead to increased demand for natural resources and growing waste streams.

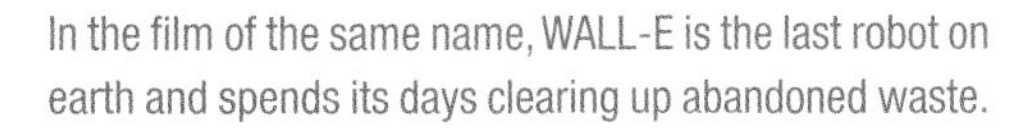

In the film of the same name, WALL-E is the last robot on earth and spends its days clearing up abandoned waste.
©Nicescene / Shutterstock.com

Robots are full of rare materials

A robot is made out of a wide range of materials. Its body is often made of steel or aluminium, whilst its electronics call for a large amount of precious metals such as copper, silver, gold and palladium. Robots that run on batteries also contain metals such as lithium, cobalt and nickel and the magnets in their motors are made using neodymium. Our planet produces only a limited supply of many of these valuable substances, so it is of the utmost importance that they are properly recycled.

This applies equally to other electronic devices such as smartphones and computers. It is estimated that unused mobile phones and smartphones around the world contain more than 140 kilos of gold and, according to Belgian waste processing organisation Recupel, only 1 to 3 percent of this is recycled. To put this in context: 1 tonne of gold ore contains on average about 3 grams of gold, whereas the recycling of 1 tonne of smartphones yields 300 grams of gold. Discarded objects and waste could therefore represent a valuable source of materials.

Regardless of the application, it is important to use resources sparingly, and robotics is no exception. During their design, robots are optimised to achieve the best possible performance with as little material as possible. Lighter materials, such as plastics, are often used, reducing both the robot's weight and its energy consumption. These plastics or polymers are synthetic products originating from the chemical processing of petroleum products.

Circular economy

To create robots, raw materials need to be mined and processed. Most of these resources – for example ores and petroleum products – are not renewable and our stocks are currently being rapidly depleted. It is therefore crucial that we use these natural resources and the materials produced from them sensibly and economically. All too often, life cycles in many applications, including robotics, are still a linear economic process. Products are manufactured, used and then disposed of as waste. This gives rise to an ongoing process of consuming new raw materials and generating ever-increasing waste flows. A better approach would be to reuse the components or materials from discarded robots. This is known as a 'circular economy' approach, in which no materials are lost, meaning that we don't have to keep mining new ones. Material resources remain in the economy for as long as possible.

In practice, this can be achieved in several ways, with the use of renewable resources being one important solution. Discarded robots or other industrial devices represent a source of renewable raw materials that can be converted into new products. This is the largest possible cycle and means that waste is given a new life cycle and the mining of new raw materials can be reduced or even prevented. Shorter cycles include the repair and reuse of existing robots so that they can be kept in service for longer. Dismantlability for repairs or for the reuse of components, is another efficient way of repurposing parts of robots.

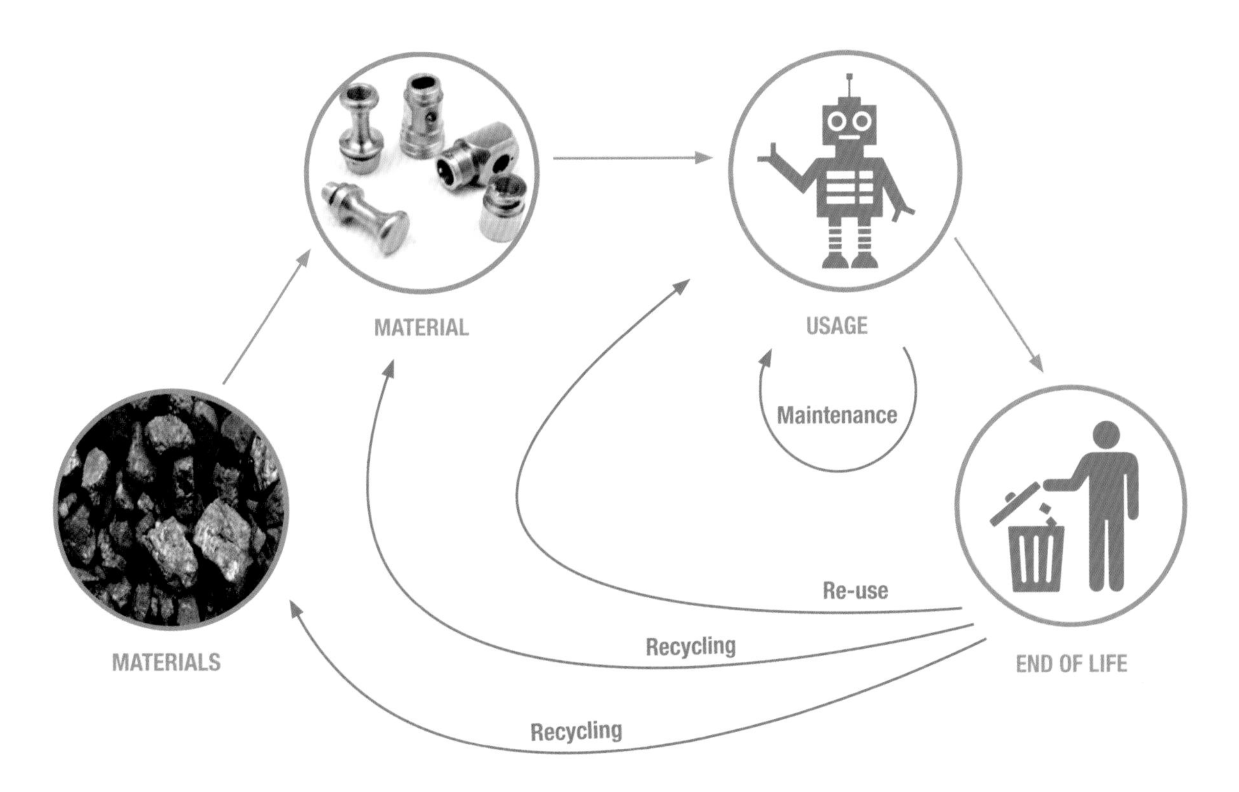

Approach to a circular economy

Researchers often take inspiration from nature when developing new materials or designing robotic components. These developments are carried out for a whole range of applications and purposes.

Recyclable cross-linked polymers

Not all the plastics or polymers used to make robotic components are recyclable. Conventional cross-linked polymers such as rubbers and thermosets can only be moulded into shape once. Once they have cured, they can no longer be reformed or reprocessed, meaning that they are non-recyclable and cannot be repaired without glue. Recently, however, it has become possible to produce cross-linked polymers with reversible chemical bonds that can be broken and reformed by the appropriate stimulus. In use, these materials exhibit similar properties to their conventional predecessors but they can take on a new form when exposed to heat or light. This means that they can be recycled or repaired.

Self-healing materials

Researchers from the Physical Chemistry and Polymer Science (FYSC) research group at the VUB have taken this one step further by developing materials that are not only recyclable, but can also repair themselves when damaged. When reversible polymer networks are overloaded, the reversible bonds are the weakest and the first to break. These bonds either reform spontaneously or can be induced to do so by applying the appropriate stimulus, such as heat or light.

The researchers are also trying to mimick the human cardiovascular system – which is responsible for the circulation of blood in our bodies – in synthetic materials. These materials contain a healing product stored in a vascular network that is released as soon as damage occurs. This healing process is similar to that which takes place in our skin or bones. When we damage our skin, blood is released: the platelets solidify and close the wound. In the same way, the healing product in the plastics seals off the damage.

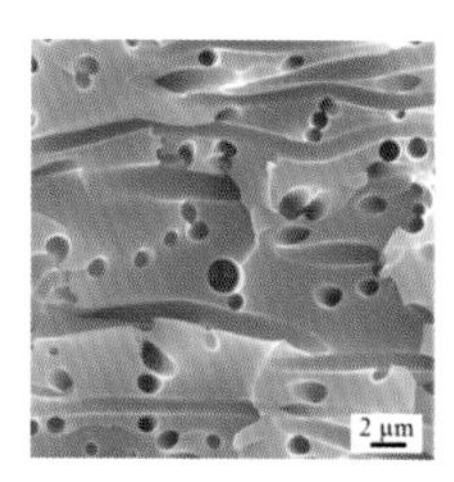

Left: a schematic vascular network.
Right: a material with a self-healing network

Artificial skin

Much research is currently being conducted into the creation of artificial skin for robots. This research draws inspiration from human skin, which is soft and flexible and protects the structures beneath it. To help robots interact with people, objects and their surroundings, the skin incorporates sensors that can detect quantities such as pressure. These sensors are based upon flexible electronics or flexible, conductive nanocomposites, which can measure pressure or changes in shape so they can determine the location and extent of any deformation. This allows the robot to sense when it is in contact with an object, and protect itself and its surroundings from damage. This principle can also be used to establish the location of any damage so that repairs can be targeted appropriately. The skin can also be made self-healing using the materials described above.

Soft robots

Researchers are also drawing inspiration from the special characteristics of squids and octopuses. Because they do not have a hard skeleton, the bodies of these animals are very flexible and adaptable, allowing them to take on forms and reach places that would otherwise be impossible. In a similar way, soft robots can be designed that are mainly constructed from flexible materials, allowing for safer interaction with human beings and objects. The most famous example is probably Baymax, the lovable robot from the Walt Disney film *Big Hero 6*.

However, the soft materials from which these robots are made, are are susceptible to damage by sharp objects. For this reason, research is carried out into materials that can self-heal so that the life span of these robots can be significantly extended.

Researchers from VUB's Brubotics lab developed a grab, an artificial muscle and an artificial hand made of self-healing materials. These materials are able to repair any damage suffered so the hand can resume its duties after the healing process.

A robotic hand made of self-healing materials from the VUB Brubotics lab.

Advanced production techniques

Advanced production techniques are needed to produce robotic components of increasing complexity. One approach that is currently experiencing a phenomenal upsurge is 3D printing, a technique that allows complex shapes to be produced with a very limited amount of material waste. Moreover, waste from recyclable materials can easily be reused. If you want to produce a single, highly specific object, 3D printing is much more versatile than fixed production lines and allows designs to be rapidly adjusted before the manufacture of the parts.

Bio-based and biodegradable materials

But what if a discarded robot could simply be thrown on the compost heap? In addition to recyclable materials, a wide range of biodegradable materials are being developed. These can be broken down organically when they have reached the end of their service life, preventing waste from accumulating in landfill or, worse still, finding its way into nature or our oceans. The current trend in robotics is to make robots stronger and increase their lifespans, but that does not necessarily rule out biodegradability. Biodegradable materials can last a long time when not exposed to the specific bacteria that break them down. It is also possible to make products from raw materials that are both organic and renewable.

Can robots themselves do their bit towards sustainability?

Robots are already being used to sort waste in recycling centres, where they separate out different types of plastic or non-recyclable waste, performing this task much faster and more efficiently than human beings. They can also be used to clear up and collect waste. In the Walt Disney film *WALL-E*, robots are used to collect and compress the mountains of waste, that human beings have left behind and, in real life, Apple has developed the robot Daisy that can dismantle and recycle 200 iPhones per hour. The extracted materials are used to produce new iPhones. The same principle can also be applied to robots, which are already being made by other robots. Robots can also be used to dismantle other robots and sort the parts and materials for reuse and recycling.

It is also possible to deploy robots to search for specific substances, and they have the advantage that they can perform such tasks even in locations that are hazardous or inaccessible to human beings. For example, there are robots that mine precious metals in difficult-to-reach locations like the bottom of the ocean.

Robots can also be used to monitor and study our surroundings, the environment and nature, often working in locations, or at times, that would not be feasible for humans. For example, they can monitor pollution or the quality of our environment, or map the harmful consequences of climate change. But the impact that these robots have on the environment they come into contact with must always be taken into account.

Waste robots instead of robotic waste?

More robots mean more materials to produce them. If we are not very careful about how we use these materials, yet more waste will be created during the production and use of robots and at the end of their life times. However, it is possible to increase the sustainability of robots, by increasing their lifespans and by recycling materials, components or even entire robots. Robots could also reduce the waste mountain by getting involved in collection and recycling. So perhaps we will soon have a waste robot in our homes helping us collect, separate and sort our rubbish – thereby increasing recycling rates – rather than a garbage bag for robotic waste.

ROBOTS CAN LEND A HAND

Robots can be put to work to do jobs that increase the sustainability of our society and protect the environment, and are already making themselves useful in the collection, sorting and processing of household waste. But they can also be used to deal with chemical spillages – for example when a truck sheds its load of chemicals or in the event of an oil leak from a drilling platform – or to process nuclear waste. In addition, they can be used to check installations to prevent accidents.

Robots are also used to monitor our environment, helping us to map out the quality of the world we live in. They can test air, water and soil quality, predict, or detect the first signs of, environmental disasters such as volcanoes, tsunamis and hurricanes, and map the effects of global warming.

OceanOne, developed by Stanford Robotics Lab, assists human divers in their research of the La Lune shipwreck, at a depth of 100 metres in the Mediterranean Sea.

©Frederic Osada and Teddy Seguin, DRASSM/Stanford University

SO WHAT'S TO BE DONE?

HOMO ROBOTICUS – THE SYMBIOSIS BETWEEN MAN AND TECHNOLOGY

As Darwin was well aware, it's not the strongest animal that survives, but the one that is best fit for its environment. This means we will have to continue adapting to the emerging trend of robotization. Let us not underestimate the capabilities of mankind. We have millions of years of evolution behind us, evolution which robots are yet to undergo. Both cognitively and physically, the human 'machine' is much more impressive and complex than its robotic counterpart. Yet our bodies also have their limitations and we face social challenges.

Moreover, there will soon be more artificial agents – both physical robots and digital bots – on the planet than human beings, and they will possess a degree of complexity that we cannot yet fully comprehend. Algorithms will come to know certain aspects of our lives better than we do ourselves. So we can't afford to rest on our laurels. As Henri Poincaré argued: 'Thought must never submit, because this would mean the end of thinking.'

The homo roboticus does not submit to robotics and artificial intelligence, instead using technology for the benefit of his or her life, work and society. He or she focuses on developing and deploying technology that is complementary to human beings, working with technology, but remaining in charge and continuing to bear ultimate responsibility. To achieve this, he or she creates an ethical and legal framework that evolves proactively along with changing technological possibilities. He or she considers freedom, solidarity and equality to be of paramount importance, and adapts technology to the needs of mankind by co-creation, using a multidisciplinary approach. He or she ensures that everybody has a meaningful place in society. This allows homo roboticus to spend time with others and engage in mutual care and support, thereby becoming more fully human.

Humane recommendations for the technological society

In this book, academics from various disciplines have investigated the place of man in the technological society of the future. There was a valid reason behind the investigation of this topic. In the past, scientific and technological developments have radically altered the world in which we live. This process will continue into the twenty-first century. Technology will also be an important aspect of the solution to current and future societal challenges, such as the ageing population, rising health costs, the need for healthier and better work, the environment, climate change, mobility, the depletion of raw materials and the issue of energy resources.

But the growing role of technology in tackling these social problems also gives rise to a new and very important challenge: How can we continue to ensure that technology benefits human beings and society? How can we co-evolve the social and technological changes towards a form of robotization that is beneficial, rather than detrimental, to mankind? If we want to maintain mankind's central position in the technological society of the future, there are a number of steps that we urgently need to take. The authors of this book have gathered steps into an inclusive robot agenda, with ten specific recommendations.

The ten recommendations for homo roboticus start from a position of radical humanism. They place technology at the service of man and society. They show how the principles of freedom, equality and solidarity can be upheld and consolidated. These recommendations are designed to establish a path for policy and for participants in our society, so that together we can derive the maximum benefit from technology, whilst keeping its negative aspects to a minimum. Our future is in our own hands. This initiative is a call for action and evolution. A call to the homo roboticus.

1. Raise awareness about technology and encourage open innovation

There is a lot of fear surrounding the issue of technology. In 2017, nearly nine out of every ten Europeans were convinced that robotics and artificial intelligence need careful management. As in the history of every disruptive innovation, people are concerned about the consequences of the changes that they are seeing around them. Moreover, the innovations are coming at us thick and fast. This gives rise to a feeling of powerlessness in terms of influencing the future. At the same time, people are also curious about the potential of new applications. For example, 68 percent of Europeans believe that robots and artificial intelligence are good for society because they can help people do their jobs and carry out day-to-day household chores. So there is something of a paradox here.

This shows that what we need is technological awareness and a public debate about the part that we want robots to play in the future. This kind of debate calls for voices and visions from different domains, or the famous '*quadruple helix*' model for open innovation, which brings together government, industry, universities and citizens. The voices of artists also need to be heard, so that we can benefit from their sense of aesthetics and critical reflection on the world. By bringing together multidisciplinary visions in an open ecosystem, new forms of added value can be created that benefit multiple parties.

2. Encourage technological co-creation

Technological choices are made by human beings, and we need to make sure that such acts are undertaken in a thoughtful and democratic way. Technical disciplines therefore need to cooperate with researchers from the social, humane and medical sectors. End-users also need to be part of the development process right from the outset, so that technological solutions are not simply imposed upon them. The search for added value must not be allowed to take place without consideration for the new conflicts and disadvantages that this may involve. But we can focus primarily on win-win situations. At the same time, we need to keep one eye on the unequal distribution of the joys and the burdens that may arise.

Different kinds of people, groups and organizations need to be included in that debate. Groups that are not automatically inclined to participate in a debate on technology should be actively sought out and drawn in from different organizations, including the government. This can be done by offering targeted information and highlighting the topics that affect these groups. After all, this is a subject that concerns everyone: at home, on the street, and at work.

During the Homo Roboticus launch event, Viktoria Modesta, the first bionic pop artist, said: 'One of the saddest things is when people think that thinking about the future isn't for them. Everybody should be engaged with how the future of their life and their environment is going to be.'
©Jean Cosyn - VUB

3. Redistribute work between man and technology

One question that we will have to face is: *Which jobs and responsibilities do we want people to hang on to, and which do we prefer to leave to robots?* In many cases, it is ultimately the customer who will decide. Will people opt for self-driving cars or would they rather keep their hands on the wheel? Will they eat at robot restaurants – which are already making a big splash in several cities – or will they stick with their favourite bartender or waiter? In the future, will they shop exclusively online or continue to enjoy visiting our many town centres? Even the experts can't yet provide clear answers to these questions.

We all know that many aspects of our jobs, especially repetitive office jobs, can be taken over by technology. But it is still unclear whether this will actually happen on a large scale. Many experts point out that not only will some jobs disappear, new – as yet unheard of – ones will also be created, such as *chief trust officer, data detective* or *man-machine teaming manager*.

However, everyone is in agreement that the way we do many of our jobs will change as a result of technology. In China, for example, people are already experimenting with artificial teachers who analyse essays, assist with pronunciation, and guide students who are having difficulties with mathematics. So will teaching staff become superfluous? The answer is no. Teachers inspire children and train them to ask questions, distinguish between information sources, weigh up consequences and, in particular, to be creative. The robot is a classroom tool that supports teachers: it is not going to take over their jobs. Human teachers are still an essential part of the education of all our children.

The central position of human beings and the supporting role of technology is a choice that we, as a society, must also apply to care, where we still have people looking after other people. Our ageing population means that this is another field in which robots will have a role to play, but human beings will still act as healthcare providers.

Although a robot is strongly inspired by the human body and brain, humans and robots go about things in fundamentally different ways. Our strengths are complementary. Robots can help with boring, precise and heavy work, allowing human beings to focus on helping each other, creativity, dexterity and problem solving.

We are therefore in favour of investing in and developing complementary machines that allow human beings and robots to work together, rather than substituting robots for people. The goal of our economy should not be to replace as many people as possible with robots, but to improve production by getting human beings and robotics to work together, making our jobs healthier and more attractive.

4. Support lifelong learning

In the 2017 Eurobarometer survey, 74 percent of Europeans thought that the use of robots and artificial intelligence would cause more jobs to disappear than new ones to be created. Yet, as many as 53 percent of EU citizens in employment believed that their own jobs could not be done – even partially – by a robot or artificial intelligence. So we think it's going to happen to other people, but not to ourselves. We know from various studies that artificial intelligence and robotics will indeed have a substantial impact upon employment. Not only will certain jobs disappear as a result of technology, the content of our jobs will also change. Almost all jobs will increasingly require employees to learn to work with technology.

This calls for ongoing training in new technologies, starting in school and continuing into the workplace. Young people need to be taught to excel in those areas where robots are weak, such as creativity, empathy and cooperation. In industry, people with a background in social sciences, philosophy or linguistics are already highly sought after to help develop robots and artificial intelligence in a man-made world. In a world surrounded by technology, STEM-(science, technology, engineering, maths) and computational thinking – formulating or reformulating problems in such a way that they can be solved using computer technology – will have to be given much greater emphasis in the attainment targets of compulsory education.

But, in a rapidly changing world, learning does not stop when schooling finishes. To prevent employees from missing the boat in this area, the government must continue to encourage and support lifelong learning in the workplace. Universities and colleges have an important role to play here. They need to adapt the range of courses on offer to lifelong learning through measures such as the organization of short-term courses in the evenings or at weekends. It goes without saying that universities and further education colleges need to be given the necessary public support.

5. Encourage technological entrepreneurship

Robots and artificial intelligence represent an immense new market. This will, in turn, give rise to new services and ecosystems that may be even more significant. The size of these markets means that robots are in a position to disrupt and transform existing economies. Although there are certainly successful European ICT companies, they do not have the global visibility that American and Asian companies enjoy. Can we prevent the same scenario from playing out in robotics? Can European robotics companies help us become not only a consumer but also a producer of robots? This would bring about new prosperity and jobs, and mean that companies incorporated European values and standards into their products.

Investing in robotics at this time is a tricky business, but the prototypes from the research labs still need to evolve towards market-readiness. However, technology and markets still need to be developed further. This will take time and money. Who in Europe is prepared to take these great risks? If we wait too long before founding companies, we run the risk that monopolies will be established and rule our lives, as the GAFA companies – Google, Apple, Facebook and Amazon – do now. It is not only American companies that threaten to gain the upper hand; China has also woken up and wants to dominate the world with its robotics and artificial intelligence.

So there is an urgent need for more, and more successful, Belgian and European technology companies. Governments are often accused of blocking technology and innovation, but these same governments – in Europe in particular – are the biggest risk investors in research and development. Today's commercial technological products have their origins in basic – mostly government-funded – research, which started out in laboratories decades ago. This basic research is always needed in robotics and artificial intelligence.

Legislation must also follow technological developments so that innovative companies have the legal certainty they need to grow without infringing the fundamental rights of citizens. To meet these requirements, the government needs to allocate resources to the support of technological innovation and provision of the necessary training.

6. Reform the funding model for the welfare state

Since 1890, Belgians have celebrated Labour Day on 1 May to commemorate the battle for the eight-hour working day. Just as at the beginning of the last century workers had to adapt to industrial assembly lines, today's workers have to adapt to technological assembly lines. We need to adapt our labour market to this new technological world: it goes without saying that we cannot leave workers to their technological fate. As a society, we must take steps to respond to this evolution, so that everyone can participate and have a role in society.

We also need to take another look at how we are going to continue financing the welfare state, something that is currently achieved on the basis of contributions from human labour. If we want to guarantee the funding of the welfare state as we currently know it into the future, we will have to think about an alternative financing model to deal with the budgetary consequences of work-related technology. This can be done, firstly, by opting for a financing model that is less dependent on labour and, secondly, by ensuring that the added value created by technology supports financing without slowing down innovation.

7. Require technological developments to be sustainable

The United Nations' *Sustainable Development Goals* are the world's goals for sustainable development for 2030. The seventeen main objectives cover areas such as climate, security, education, renewable energy and poverty.

Robotics and artificial intelligence have the potential to speed up progress towards these developmental goals, but also bring with them complex challenges. For example, the *World Food Program* is looking into self-driving trucks for transporting food in war zones, and drones with sensors that can perform AI analyses have the potential to increase agricultural productivity (almost 50 percent of crops are lost through waste, over-consumption and production inefficiencies).

On the other hand, studies suggest that the burden of job losses due to robotization is likely to be primarily borne by low-wage countries, as the traditional advantage of low labour costs is eroded. It will also be much more difficult for these countries to put in place social measures such as improved education and redistribution models due to their much smaller economic base.

In fact, we don't really know what the impact of robotics will be. Studies take many different directions and are often based on guesswork. Rather than predicting the future, we will have to create it for ourselves. Politicians, civil society and trade unions will need to take scientific studies as the basis for drawing up guidelines that allow robots and artificial intelligence to be used in a way that creates a sustainable world.

8. Create a charter for ethical technology

In the future, we will cooperate with robots at work and welcome them into our homes. They may even become a part of our bodies. But technology is far from neutral. Whether we know it or not, there are ethical and legal aspects to both its design and use. To a large degree, society can be created, and robotics is going to play an indisputable role in this. How do we want to live? What kind of society do we want? We are constantly making choices. Safety, responsibility and privacy are fundamental human rights, and must not be simply surrendered to robots.

Making good choices in the public interest is not an exact science and is made more complicated by the fact that different parts of society will have different preferences. That is why it is essential that robot developers and robotics companies also engage with these aspects with greater levels of awareness and, together with other experts such as philosophers, lawyers and sociologists, reflect on how they are shaping the future. We must continue to ensure that technological development reflects human values.

9. Avoid technological inequality

It is clear that technology can open up a wide range of possibilities and be of huge benefit to mankind. The opportunities that technology offers are seemingly endless and are manifested in every possible way. However, this immense potential can also give rise to inequality between those who can afford the technology and those who cannot. We must be vigilant. It is not only a question of economic inequalities between, for example, companies that have the means to use big data technology and companies that do not, but also of human inequality.

Disabled people can increasingly rely on technological assistance. Prosthetic arms and legs can replace limbs, exoskeletons can improve the mobility of people with paralysis, and technology can be controlled by capturing brain signals. Here, too, there is a danger of increasing the inequality between those who can afford this technology and those who cannot. It is up to the government to put in place measures to ensure equality of access.

10. Let robots be robots

The Geminoid HI-1 and its designer Prof Ishiguro – or is it the other way around?
©Osaka University

Although man and machine operate on fundamentally different principles, robots can hold up a mirror to humanity. Robots can be experimental platforms for validating or rejecting biological and psychological processes. At the same time, we can learn more about ourselves by investigating how we interact with our artificial counterparts.

Using robots leads us to question who we are and how we project our desires onto these creations. Theories suggest that we perceive robots through human filters based upon *nature* and *nurture*. We interact blindly with robots and other technologies as if they were human, and we see human traits in them, including thoughts and emotions. Moreover, we feel more of an affinity with real physical robots than we do with virtual robots or computers, suggesting that a physical embodiment is important in our relationships with artificial technologies. Studying human behaviour and functioning allows us to design better and smarter technology. At the same time, tests on robots allow us to gain greater awareness of human cognition, emotion and behaviour.

But what about when technology becomes increasingly indistinguishable from human beings? Google has released audio clips of its Duplex system taking restaurant and hair salon bookings over the telephone: the person on the other end didn't even realize they were talking to a machine. Filler words such as 'uh' and 'mmm-mmm' help a lot in this regard.

By opening up human communication channels such as emotions and gestures to robots, we put the focus on the human aspect and make it easier to communicate with robots. Robots that attempt to mimic the human face and human movements are even being developed. But isn't this misleading? Couldn't it lead to manipulation? To prevent this, technology needs to be identified as such from the outset when interacting with human beings.

LITERATURE

INTRODUCTION

How can robots and humans live side by side?

Manyika, J., Chui, M., Bughin, J., Dobbs, R., Bisson, P., & Marrs, A. (2013). *Disruptive technologies: Advances that will transform life, business, and the global economy* (Vol. 180). San Francisco, CA: McKinsey Global Institute.

Christensen, H.I. et al. (2013). *A roadmap for US robotics: from internet to robotics.* Robotics Virtual Organization.

Siciliano, B., & Khatib, O. (Eds.). (2016). *Springer handbook of robotics.* Springer.

UTOPIA/DYSTOPIA

What can science fiction contribute to the field of humanistic robotics?

Asimov, I. (1954). *The Caves of Steel.* New York: Bantam Books.

Asimov, I. (1957). *The Naked Sun.* New York: Bantam Books. Asimov, I. (1981). *Asimov on Science Fiction.* Garden City, New York: Doubleday & Company.

Asimov, I. (1983). *The Robots of the Dawn.* New York: Bantam Books.

Asimov, I. (1985). *Robots and Empire.* New York: Doubleday Books.

Asimov, I. (1995). *The Complete Robot.* London: Harper Collins.

Čapek, K. (2001, transl. Paul Selver and Nigel Playfair). R.U.R. New York:
Dover Publications.

Cave, S., & Dihal, K. (2018). The automaton chronicles. Our responses to robots echo down the millennia. *Nature 559,* 473-475.

Clute, J., & Nicholls, P. (1999). *New Encyclopedia of Science Fiction.* London: Orbit books.

De Bodard, A. (2018). *The Teamaster and the Detective.* Jabberwocky Literary Agency, Inc. (e-book).

Dick, P.K. (2010). *Do androids dream of electric sheep.* London: Gollancz.

Gibson, W. (1984). *Neuromancer.* New York: Ace Science Fiction.

Gibson, W. (1987). *Count Zero.* London: Grafton Books.

Gibson, W. (1989). *Mona Lisa Overdrive.* London: Grafton Books.

Gunkel, D.J. (2018). *Robot Rights.* Cambridge, Massachusetts: MIT Press.

Heinlein, R. A. (1959). Science Fiction: It's Nature Faults and Virtues. In *The Science Fiction Novel: Imagination and Social Criticism* (pp. 14-48). Chicago: Advent Publishers. Retrieved from http://sciencefiction.loa.org/biographies/heinlein_science.php

Herbert, F. (1965) *Dune.* New York: Penguin Random House.

Holmes, R. (2016). The Science that fed Frankenstein. *Nature* 535, 490-492.

Leckie, A. (2013). *Ancillary Justice.* London: Orbit.

Melville, A.T. (2014). *Android Rebellion. The Future is a paradox.* S.l.: CreateSpace Independent Publishing Platform.

Pappas, S. (2016). *Collecting Science Fiction - Karel Čapek and the Origin of the Word Robot.* Retrieved from https://www.ilab.org/articles/collecting-science-fiction-karel-capek-and-origin-word-robot

Samuelson, D.N. (1993). Modes of Extrapolation: The Formulas of Hard SF. *Science Fiction Studies, 20*(2), 191-232.

Shelley, M.W. (1818). *Frankenstein or the Modern Prometheus.* London: Lackington, Hughes, Harding, Mavor & Jones.

Tchaikovsky, A. (2017). *Dogs of War.* London: Head of Zeus.

Watts, P. (2014). *Firefall.* London: Head of Zeus.

Westfahl, G. (2015). The Mightiest Machine: The Development of American Science Fiction from the 1920s to the 1960s. In G. Canavan & E.C. Link (Ed.), *The Cambridge companion to American Science Fiction* (pp. 17-30). Cambridge: Cambridge University Press.

HUMANOID

Human or robot: which makes the best machine?

Cheetah robot 'runs faster than Usain Bolt. BBC News (2012, September 6). Retrieved from https://www.bbc.co.uk/news/technology-19506130.

Madden, J.D. (2007). Mobile Robots: Motor Challenges and Materials Solutions. *Science 318,* 1094-1097.

Will a robot be your best friend?

Anki. (2018, August 8). *Vector by Anki. A Giant Roll Forward for Robotkind* [Online video]. Retrieved from https://www.youtube.com/watch?v=Qy2Z2TWAt6A

Bauman, Z. (2001). *Community: Seeking Safety in an Insecure World.* Cambridge: Polity.

Beck, U. (1994). *Riskante Freiheiten: Individualisierung in Modernen Gesellschaften.* Frankfurt: Suhrkamp Verlag.

Bourdieu, P. (2013). *Distinction: A Social Critique of the Judgement of Taste.* London: Routledge.

Breazeal, C. (2004). *Designing Sociable Robots.* Cambridge, Massachusetts: MIT Press.

Cacioppo, J.T., & Patrick, W. (2008). *Loneliness: Human Nature and the Need for Social Connection.* New York: Norton.

Dautenhahn, K. (2007). Socially Intelligent Robots: Dimensions of Human–Robot Interaction. *Philosophical Transactions of the Royal Society of London B: Biological Sciences,* 362 (1480), 679-704.

Durkheim, E. (1984). *The Division of Labor in Society.* New York: Free Press.

Luhmann, N. (1995). *Social Systems.* Palo Alto: Stanford University Press.

Mori, M. (1970). The Uncanny Valley. *Energy* 7(4), 33–35.

Putnam, R. (2001). *Bowling Alone: The Collapse and Revival of American Community.* New York: Simon & Schuster.

Robot Mario, the new employee and mascot of the Marriott Hotel Ghent. (2015, August 18). Retrieved from https://www.horecatrends.com/en/robot-mario-the-new-employee-and-mascot-of-the-marriott-hotel-ghent/

Simmel, G. (2012). The Metropolis and Mental Life. In J. Lin & C. Mele (Ed.) *The Urban Sociology Reader* (pp. 37–45). London:Routledge.

Schirmer, W., & Michailakis, D. (2018). Inclusion/Exclusion as the Missing Link. A Luhmannian Analysis of Loneliness Among Older People. *Systems Research and Behavioral Science* 35(1), 76–89.

Turkle, S. (2011). *Alone Together: Why We Expect More from Technology and Less from Each Other.* New York: Basic Books.

Van Dijck, J. (2013). *The Culture of Connectivity: A Critical History of Social Media.* Oxford: Oxford University Press.

Weber, J. (2013). Opacity versus Computational Reflection. Modelling Human-Robot Interaction in Personal Service Robotics. *Science, Technology & Innovation Studies* 10(1), 187–99.

Weizenbaum, J. (1966). ELIZA - a Computer Program for the Study of Natural Language Communication between Man and Machine. *Communications of the ACM* 9(1), 36–45.

Zhao, S. (2006). Humanoid Social Robots as a Medium of Communication. *New Media & Society* 8(3), 401–19.

Will you fall head over heels in love with a robot?

Ferguson, A. (2010). *The sex doll: A history.* Jefferson, North Carolina: McFarland Press.

Finkel, E. J., Hui, C. M., Carswell, K. L., & Larson, G. M. (2014). The suffocation of marriage: Climbing Mount Maslow without enough oxygen. *Psychological Inquiry* 25, 1– 41.

Fowers, B. J., Laurenceau, J. Penfield, R. D. Cohen, L. M. Lang, S.F. et al. (2016). Enhancing relationship quality measurement: The development of the Relationship Flourishing Scale. *Journal of Family Psychology* 30(8), 997-1007.

Goode, W. J. (1982). *The Family.* Englewood Cliffs, New Jersey: Prentice Hall.

Levinger, G. (1983). Development and change. In Kelly, H.H. (Ed.), *Close relationships* (pp. 315-359). New York: W.H. Freeman and Company.

Levy, D. (2007). *Love and Sex with Robots.* New York: Harper Collins Publishers.

Ornella, A. D. (2015). Uncanny Intimacies: Humans and Machines in Film. In M. Hauskeller, Philbeck, T. & Carbonell, C. (Ed.), *The Palgrave Handbook of Posthumanism in Film and Television* (pp. 330-338). London: Palgrave Macmillan.

Scheutz, M., & Arnold, T. (2016). Are we ready for sex robots? *The Eleventh ACM/IEEE International Conference on Human Robot Interaction (pp. 351-358).* Plaats uitgave: uitgeverij.

Scheutz, M., Arnold, T. (2017). Intimacy, bonding, and sex robots: Examining empirical results and exploring ethical ramifications. In J. Danaher & N. McArthur (Ed.), *Robot Sex: Social and Ethical Implications* (werktitel). Cambridge, Massachusetts: MIT Press.

Sharkey, N., van Wynsberghe, A., Robbins, S., & Hancock, E. (2017). *Our Sexual Future with Robots.* Retrieved from http://responsiblerobotics.org/wp-content/uploads/2017/07/FRR-Consultation-Report-Our-Sexual-Future-with-robots_Final.pdf

META

How illogical do robots have to be for us to live with them?

Van Bendegem, J.P. (2011). *Logica.* Antwerp: Luster.

Batens, D. (2017). *Logicaboek.* Antwerp: Garant.

Kahneman, D. (2011a). *Ons feilbare denken.* Amsterdam/Antwerp: Contact.

Kahneman, D. (2011b). *Thinking, Fast and Slow*. New York: Farrar, Straus & Giroux.

Do robots have a body and a mind too?

Brooks, R.A. (1991). Intelligence without reason. In R. Chrisley & S. Begeer (Ed.), *Artificial intelligence. Critical concepts* (pp. 107-163). London and New York: Routlegde

Kandinsky, W. (1977). *Concerning the Spiritual in art.* New York: Dover Publications.

Do robot designers have prejudices?

Campaign to Stop Killer Robots. Retrieved from https://www.stopkillerrobots.org

Campaign Against Sex Robots. Retrieved from https://campaignagainstsexrobots.org

Moral Machine. Human Perspectives on Machine Ethics. Retrieved from http://moralmachine.mit.edu

Richardson, K. (2015). The Asymmetrical 'Relationship': Parallels between Prostitution and the Development of Sex Robots. *ACM SIGCAS Computers and Society. Special Issue on Ethicomp 45*(3), 290-293.

Schwab, K. (2017). Nest Founder: I Wake Up in Cold Sweats Thinking, What Did We Bring to the World?, *Fastcompany 17*. Retrieved from https://www.fastcompany.com/90132364/nest-founder-i-wake-up-in-cold-sweats-thinking-what-did-we-bring-to-the-world

van de Poel, I., & Royakkers, L. (2011). *Ethics, Technology, and Engineering: An Introduction*. London: Wiley-Blackwell.

Weizenbaum, J., & Wendt, G. (2006/2015, transl. Benjamin Fasching-Gray). *Islands in the Cyberstream. Seeking Havens of Reason in a Programmed Society*. Sacramento, Californië: Litwin Books.

Zierse, M. (2017, July 25). Fabrikant van slimme stofzuigers wil de plattegrond van uw huis doorverkopen. *Trouw*. Retrieved from https://www.trouw.nl/home/fabrikant-van-slimme-stofzuigers-wil-de-plattegrond-van-uw-huis-doorverkopen~acd9e66d/

AUTONOMY

Which country will seize power over robots?

Antipater van Thessaloniki (1925, transl., W. R. Paton). *Epigrams*. Cambridge, Massachussets: Loeb.

Who is Winning the AI Race. *MIT Technology Review* (2017, June 20). Retrieved from https://www.technologyreview.com/s/608112/who-is-winning-the-ai-race/

Bland, B. (2016, June 6). China's robot revolution. *Financial Times*. Retrieved from https://www.ft.com/content/1dbd8c60-0cc6-11e6-ad80-67655613c2d6

Bourguignon, F., & Morrison, C. (2002). Inequality among world citizens: 1890-1992. *American Economic Review* 92(4), 727–744.

Breitzman, A. & Thomas, P. (2017). Patent Power 2017, IEEE Spectrum.

Bryant, C., & He, E. (2017, January 8). The Robot Rampage. Bloomberg. Retrieved from https://www.bloomberg.com/opinion/articles/2017-01-09/the-robot-threat-donald-trump-isn-t-talking-abou

International robot statistics, International Federation of Robotics. Retrieved from www.irf.org

Chinezen azen op Belgische robotbouwer Robojob (2017, January 25). Retrieved from https://derijkstebelgen.be/nieuws/chinezen-azen-op-belgische-robotbouwer-robojob

European Commission (2018). *USA-China-EU plans for AI: where do we stand?* Brussels: European Commission.

European Parliament, DG for Internal Politics (2017). *European Civil Law Rules in Robotics*. Brussels: European Parliament.

Frank, M., et al. (2018). Small cities face greater impact from automation. *Journal of the Royal Society Interface 15: 20170946*. Retrieved from https://dam-prod.media.mit.edu/x/2018/02/13/20170946.full.pdf

Holslag, J. (2018). *A Political History of the World*. London: Penguin.

Homeros, (2003, transl. E.V. Rieu). *The Odyssey*. London: Penguin.

ILO (2018). *The impact of technology on the quality and quantity of jobs*. Genève: ILO.

Maddisson, A. (2008). Statistics on World Population, GDP and Per Capita GDP, 1-2008 AD. Retrieved from http://ghdx.healthdata.org/record/statistics-world-population-gdp-and-capita-gdp-1-2008-ad

Patrizio, A. (2018, April 10). Top 25 Artificial Intelligence Companies. *Datamation*. Retrieved from https://www.datamation.com/applications/top-25-artificial-intelligence-companies.html

Rapp, N., & O'Keefe, B. (2018, January 8). These 100 Companies Are Leading the Way in A.I. Fortune. Retrieved from http://fortune.com/go/venture/artificial-intelligence-ai-companies-invest-startups/

RT Staff (2017, March 21). Top 50 Robotics Companies. *Robotics Business Review*. Retrieved from https://www.roboticsbusinessreview.com/rbr/top_50_robotics_companies_of_2017/

Simonite, T. (2016, May 13). Moore's Law is Dead: Now What? *MIT Technology Review.* Retrieved from https://www.technologyreview.com/s/601441/moores-law-is-dead-now-what/

Will robots be given a license to kill?

Marshall, S.L.A. (2000). *Men against fire: The problem of battle command.* University of Oklahoma Press.

Sharkey, N. (2010). Saying 'no!' to lethal autonomous targeting. *Journal of Military Ethics* 9(4), 369-383.

Surgeon General's Office (2006, November 17). *Mental Health Advisory Team (MHAT) IV Operation Iraqi Freedom 05-07, Final Report.*

Singer, P.W. (2009). *Wired for war: The robotics revolution and conflict in the 21st century.* London: Penguin.

How do we stay in control of billions of ai agents?

Beuls, K., & Steels, L. (2013). Agent-Based Models of Strategies for the Emergence and Evolution of Grammatical Agreement. *PLoS ONE* 8(3), e58960.

Steels, L. (Ed.). (2012). *Experiments in Cultural Language Evolution.* Amsterdam: John Benjamins Publishing.

Steels, L. & Brooks, R. (1994). *The 'artificial life' route to 'artificial intelligence'. Building Situated Embodied Agents.* New Haven: Lawrence Erlbaum Ass.

Will we soon be sending our children to their sports club alone in a self-driving car?

Fagnant, D., & Kockelman, K. (2015). Preparing a nation for autonomous vehicles: opportunities, barriers and policy recommendations. *Transportation Research Part A: Policy and Practice* 77, 167-181.

Feys, M., & Vanhaverbeke, L. (2017). Consumer Acceptance of Autonomous Vehicles: A Pilot Project. In *Proceedings Automated Vehicle Symposium 2017.*

Harb, M., Xiao, Y., Circella, G., Mokhtarian, P., & Walker, J. (2018). *Projecting Travelers into a World of Self-Driving Cars: Naturalistic Experiment for Travel Behavior Implications.* TRB Annual Conference. Washington.

Lequeux, Q. (2018). *Statistisch Rapport 2017 Verkeersongevallen.* Retrieved from https://www.vias.be/publications/Statistisch%20Rapport%202017%20-%20Verkeersongevallen/Statistisch_Rapport_2017_-_Verkeersongevallen.pdf

Macharis, C., & Késuru, I. *Car sharing in Brussels: status report.* Brussels.

Road Safety: Data show improvements in 2017 but renewed efforts are needed for further substantial progress. Press release European Commission (2018, April 10). Retrieved from http://europa.eu/rapid/press-release_IP-18-2761_en.htm

Himachar, T. (2018) *PlugMyCar. A decarbonised shared and autonomous mobility solution.* Presentatie Tractebel, Brussels.

VIAS (2018). *Verkeersveiligheidsbarometer 2017.* Brussels. Retrieved from https://www.vias.be/storage/main/verkeersveiligheidsbarometer-het-jaar-2017.pdf

Viegas, J., & Martinez, L. (2016). Shared Mobility. Innovation for Liveable Cities. International Transport Forum. Retrieved from https://www.itf-oecd.org/shared-mobility-innovation-liveable-cities

Viegas, J., & Martinez, L. (2017). Transition to Shared Mobility. International Transport Forum. Retrieved from https://www.itf-oecd.org/transition-shared-mobility

Wadud, Z., MacKenzie, D. & Leiby, P. (2016). Help or hindrance? The travel, energy and carbon impacts of highly automated vehicles. *Transportation Research Part A: Policy and Practice* 86, 1-18.

WHO (2018). *Key Facts Road Traffic Injuries.* Retrieved from http://www.who.int/en/news-room/fact-sheets/detail/road-traffic-injuries.

WORK

Will robots take our jobs?

Acemoglu, D., & Restrepo, P. (2017). Robots and jobs: Evidence from the US. *VOX: CEPR Policy Portal.* Retrieved from https://voxeu.org/article/robots-and-jobs-evidence-us

Acemoglu, D., & Restrepo, P. (2018). The race between man and machine: Implications of technology for growth, factor shares, and employment. *American Economic Review* 108(6), 1488–1542.

Autor, D.H. (2015). Why are there still so many jobs? The history and future of workplace automation. *Journal of Economic Perspectives* 29(3), 3-30.

Brynjolfsson, E., & McAfee, A. (2014). *The Second Machine Age: Work, Progress, and Prosperity in a Time of Brilliant Technologies.* New York: W.W. Norton & Company.

Chase, S. (1929). *Men and machines.* New York: Macmillan Publishers.

Cueni, R. (2017). Robots will take all our jobs. In B.S. Frey & D. Iselin (Ed.), *Economic Ideas You Should Forget* (pp. 35-36). Berlin: Springer.

Ford, M. (2016). *Rise of the Robots: Technology and the Threat of a Jobless Future.* New York: Basic Books.

Goos, M., Manning, A., & Salomons, A. (2014). Explaining job polarization: Routine-biased technological change and offshoring. *American Economic Review 104*(8), 2509-2526.

Gordon, R. J. (2016). *Rise and Fall of American Growth: The U.S. Standard of Living since the Civil War.* Princeton: Princeton University Press.

Harbour, R. & Schmidt, J. (2018, May 11). Tomorrow's Factories Will Need Better Processes, Not Just Better Robots. *Harvard Business Review.* Retrieved from https://hbr.org/2018/05/tomorrows-factories-will-need-better-processes-not-just-better-robots

International Federation of Robotics (2017). *World Robotics Report 2016.* Retrieved from https://ifr.org/ifr-press-releases/news/world-robotics-report-2016

Keynes, J.M. (1978). Economic possibilities for our grandchildren. In E. Johnson & D. Moggridge (Ed.). *The Collected Writings of John Maynard Keynes. Volume 9: Essays in Persuasion* (pp. 321–332). Cambridge: Cambridge University Press for the Royal Economic Society.

McCloskey, D., & Klamer, A. (1995). One Quarter of GDP is Persuasion. *American Economic Review 85*(2), 191-195.

Remember the mane: As robots encroach on human work, study the fate of the horse (2017, April 1). *The Economist*, 66.

Siciliano, L. (2017, May 30). Tennis balls are made using this 11-step process [Online video]. Retrieved from https://www.thisisinsider.com/how-tennis-balls-are-made-processed-penn-manufactured-satisfying-machines-2017-5

Will you end up working side-by-side with a robot?

Amandels, S., Eyndt H.O., Daenen, L., & Hermans, V. (2019). Introduction and Testing of a Passive Exoskeleton in an Industrial Working Environment. In S. Bagnara, R. Tartaglia, S. Albolino, T. Alexander & Y. Fujita (Ed.). *Proceedings of the 20th Congress of the International Ergonomics Association* (IEA 2018).

Arntz, M., Gregory, T. & Zierahn, U. (2016). The Risk of Automation for Jobs in OECD Countries: A Comparative Analysis. *OECD Social, Employment and Migration Working Papers 189.* Parijs: OECD Publishing. Retrieved from http://dx.doi.org/10.1787/5jlz9h56dvq7-en

De Looze, M.P., Bosch, T., Krause, F., Stadler, K.S., & O'Sullivan L.W. (2015). Exoskeletons for industrial application and their potential effects on physical work load. *Ergonomics 59*(5), 671-681.

Gibbs, S. (2018, April 16). Elon Musk drafts in humans after robots slow down Tesla Model 3 production. *The Guardian.* Retrieved from https://www.theguardian.com/technology/2018/apr/16/elon-musk-humans-robots-slow-down-tesla-model-3-production

Hill, D., Holloway, C.S., Morgado Ramirez, DZ, Smitham, P., & Pappas, Y. (2017). What are user perspectives of exoskeleton technology? A literature review. *International Journal of Technology Assessment in Health Care 33*(2), 160-167.

OECD Social, Employment and Migration Working Papers 189. Parijs: OECD Publishing.

Parent-Thirion, A., Biletta, I., Cabrita J., Vargas Llave O., Vermeylen G. & Wilczynska A. (2017). *Sixth European Working Conditions Survey.* Dublin: Eurofound. Retrieved from https://www.eurofound.europa.eu/surveys/european-working-conditions-surveys/sixth-european-working-conditions-survey-2015

Slatman, J. (2008). *Vreemd lichaam: over medisch ingrijpen and persoonlijke identiteit.* Amsterdam: Ambo.

Toch, M., Bambra, C., Lunau, T., van der Wel, K. A., Witvliet, M. I., Dragano, N., & Eikemo, T. A. (2014). All Part of the Job? The Contribution of the Psychosocial and Physical Work Environment to Health Inequalities in Europe and the European Health Divide. *International Journal of Health Services 44*(2), 285–305.

Is robotic shopping set to be the next big thing?

AFP (2018, January 8). The self driving store that can come to you: Toyota reveals 'roaming shops' concept that could also deliver food, packages and even offer medical services. *Daily Mail.* Retrieved from http://www.dailymail.co.uk/sciencetech/article-5247751/Toyota-brings-store-self-driving-concept-vehicle.html

Bertacchini, F., Bilotta, E., & Pantano, P. (2017). Shopping with a robotic companion. *Computers in Human Behavior 77*, 382-395.

Burgess, M. (2017, June 16). China now has a robo-grocery store that will drive to your door. *Wired.* Retrieved from https://www.wired.co.uk/article/moby-autonomous-shopping-store

De Gauquier, L., Cao, H-L., Gomez Esteban, P., De Beir, A., Van De Sanden, S., Willems, K., Brengman, M. & Vanderborght, B. (2018). Humanoid Robot Pepper at a Belgian Chocolate Shop. In *13th Annual ACM/IEEE International Conference on Human Robot Interaction* (pp. 373-373).

Eli, the first autonomous smart shopping robot [Online video]. Retrieved from https://www.robotnik.eu/portfolio/eli-the-first-autonomous-smart-shopping-robot/

Knapton, S. (2018, January 22). Fabio the robot sacked from supermarket after alarming customers. *The Telegraph.* Retrieved from https://www.telegraph.co.uk/science/2018/01/21/fabio-robot-sacked-supermarket-alarming-customers/

Roland, B. (2016). Think Act – Robots and retail – What does the future hold for people and robots in the stores of tomorrow? Parijs.

n Hoof, N. (2018, March 1). Hoe de robot zijn opmars maakt in de supermarkt. *Twinkle*. Retrieved from https://www.twinkle.be/achtergrond/170141/hoe-robot-opmars-maakt-supermarkt

HEALTH

What if athletes turn into robots?

ETH Zürich. *Cybathlon. Moving people and technology*. Retrieved from http://www.cybathlon.ethz.ch/

Erdmann, W. (2013). Problems of sport biomechanics and robotics. *International Journal of Advanced Robotic Science 10*(2), 123:130.

Press TV. *World's first ski robot challenge in S Korea* [Online video]. Retrieved from https://www.youtube.com/watch?v=6yj4S9nfgY4

Swatz L., & Watermeyer, B. (2008). Cyborg anxiety: Oskar Pistorius and the boundaries of what it means to be human. *Disability & Society* 23(2), 187-190.

Weyand, P., Bundle, M., McGowan, C., Grabowski, A., Brown, M., Kram, R. & Herr, H (2009). The fastest runner on artificial legs: different limbs, similar function? *Journal of Applied Physiology 107*, 903–911.

Do physical therapists and robots walk hand in hand?

Calabro, R.S, Cacciola A., Bertè, F., Manuli, AL, Leo, A., Bramanti, A., Naro, A., Milardi, D., Bramanti, P. (2016). Robotic gait rehabilitation and substitution devices in neurological disorders: where are we now? *Neurological Science* 37, 503-514.

Dietz, V., Nef, T., Rymer, W.Z. (2012). *Neurorehabilitation technology*. London: Springer-Verlag.

Swinnen, E., Lefeber, N., Willaert, W., De Neef, F., Bruyndonckx, L., Spooren, A., Michielsen, M., Ramon, T., Kerckhofs, E. (2017). Motivation, expectations, and usability of a driven gait orthosis in stroke patients and their therapists. *Topics in Stroke Rehabilitation*, 24(4), 299-308.

Weber, L.M. & Stein, J. (2018). The use of robots in stroke rehabilitation: a narrative review. *NeuroRehabilitation*, 43,99-110.

Do robots make good surgeons?

George, E. I., & Brand, C. T. C. (2018). Origins of Robotic Surgery: From Skepticism to Standard of Care. *JSLS: Journal of the Society of Laparoendoscopic Surgeons*, 22(4).

Nota, C. L., Smits, F. J., Woo, Y., Rinkes, I. H. B., Molenaar, I. Q., Hagendoorn, J., & Fong, Y. (2019). Robotic Developments in Cancer Surgery. *Surgical Oncology Clinics* 28(1), 89-100.

Duysburgh, P., Elprama, S. A., & Jacobs, A. (2014). Exploring the social-technological gap in telesurgery: collaboration within distributed or teams. In *Proceedings of the 17th ACM conference on Computer supported cooperative work & social computing* (pp. 1537–1548).

Tan, A., Ashrafian, H., Scott, A. J., Mason, S. E., Harling, L., Athanasiou, T., & Darzi, A. (2016). Robotic surgery: disruptive innovation or unfulfilled promise? A systematic review and meta-analysis of the first 30 years. *Surgical Endoscopy* 30(10), 4330–4352.

Can a robot put your granny to bed?

Statbel, the Belgian statistical office. Retrieved from http://statbel.fgov.be

LAW AND TAXATION

Will household robots wash your dirty linen in public?

Amazon Alexa Gone Wild!!! [Online video]. Retrieved from https://www.youtube.com/watch?v=r5p0gqCIEa8

Berbers, Y., et al. (2017). Privacy in tijden van internet, sociale netwerken and big data. *Standpunten 49*. Retrieved from http://www.kvab.be/sites/default/rest/blobs/1316/tw_privacy.pdf.

Deahl, D. (2017, July 24). *Roombas have been busy mapping our homes, and now that data could be shared.* Retrieved from https://www.theverge.com/2017/7/24/16021610/irobot-roomba-homa-map-data-sale

Kastrenakes, J. (2017, July 28). *Roomba creator says it 'will never sell your data' after talking about selling your data*. Retrieved from https://www.theverge.com/2017/7/28/16055590/roomba-wont-sell-data-irobot

Sellinger, E., & Hartzog, W. (2015, August 12). *The dangers of trusting robots.* Retrieved from http://www.bbc.com/future/story/20150812-how-to-tell-a-good-robot-from-the-bad

Van Dijck, J., Poell, T. & de Waal, M. (2016). *De platformsamenleving: strijd om publieke waarden in een online wereld*. Amsterdam: Amsterdam University Press.

Woo, M. (2014, June 5). *Robots: Can we trust them with our privacy?* Retrieved from http://www.bbc.com/future/story/20140605-the-greatest-threat-of-robots

Who's to blame – the human or the robot?

Evans, E.P. (1987). *The criminal prosecution and capital punishment of animals*. London: Faber and Faber.

Foucault, M. (1989). *Discipline, toezicht and straf.* Groningen: Historische Uitgeverij.

Girtgen, J. (2003). The historical and contemporary prosecution and punishment of animals. *Animal Law* 9, 97-133.

Tanghe, J., & De Bruyne, J. (2018). Software aan het stuur: Aansprakelijkheid voor schade veroorzaakt door autonome voertuigen. In Th. Vansweevelt & B. Weyts (Ed.), *Nieuwe risico's in het aansprakelijkheidsrecht* (pp.1-75). Antwerp: Intersentia.

Do robots have rights and obligations?

Arendt, H. (2018). *Het waagstuk van de politiek*. Utrecht: Uitgeverij Klement.

Bryson, J.J., Diamantis, M.E., & Grant, T.D. Of, for, and by the People: The Legal Lacuna of Synthetic Persons. *Artificial Intelligence and Law* 25(3), 273-291.

Dworkin, R. (1978). *Taking Rights Seriously*. Cambridge, Massachusetts: Harvard University Press.

European Parliament resolution of 16 February 2017 with recommendations to the Commission on Civil Law Rules on Robotics (2015/2103(INL).

Floridi, L. (2014). *The Fourth Revolution: How the Infosphere Is Reshaping Human Reality* Oxford: Oxford University Press.

Hildebrandt, M. (2015). *Smart Technologies and the End(s) of Law*. Abingdon: Edward Elgar.

Hildebrandt, M. (2015, December 26). Kerstessay De omgekeerde schepping. *De Standaard*. Retrieved from http://www.standaard.be/cnt/dmf20151225_02037563

Koops, B.J., Hildebrandt, M., & Jacquet-Chiffelle, D-O. (2010). Bridging the Accountability Gap: Rights for New Entities in the Information Society? *Minnesota Journal of Law Science & Technology* 11(2), 497-561.

Open letter to the European Commission Artificial Intelligence and robotics (2018). Retrieved from http://www.robotics-openletter.eu

Taxing the robots!

Frey, C., & Osborne, M. (2013). *The future of employment. How susceptible are jobs to computerisation?* Oxford Martin School. Retrieved from https://www.oxfordmartin.ox.ac.uk/downloads/academic/The_Future_of_Employment.pdf

OECD (2018). *Taxing Wages 2018*. Parijs: OECD Publishing.

SUSTAINABILITY

Mummy, can I build a robot?

Asada, M., D'Andrea, R., Birk, A., Kitano, H., & Veloso, M. (2000). *Robot in Edutainment. Proceedings of the IEEE International Conference on Robotics and Automation 2000*. San Francisco.

Johnson, J. (2003). Children, robotics, and education. *Artificial Life and Robotics* 7(1-2), 16-21.

Murphy, R.R. (2001). A Strategy for Integrating Robot Design Competitions. *IEEE Robotics & Automation Magazine 8(2), 44-55.*

Agg, Fem. (2018, August 28). Muyters wil 'wetenschapsacademie' in elke gemeente. *De Standaard*, p. 8.

Van den Berghe, W., & De Martelaere, D. (2012). Vlaamse Raad voor Wetenschap and Innovatie. Kiezen voor STEM. De keuze van jongeren voor technische and wetenschappelijke studies. *VRWI-studiereeks 25.*

Vandewalle, J., & Veretennicoff, I. (2015). *De STEM-leerkracht. Standpunten 38.* Retrieved from http://www.kvab.be/sites/default/rest/blobs/121/nw_mw_stemleerkracht.pdf

VDAB (2018). *Knelpuntberoepen in Vlaanderen 2018.* Retrieved from https://www.vdab.be/sites/web/files/doc/trends/Knelpuntberoepen_2018.pdf

So, where are those entrepreneurs in robotics?

Clarysse, B. (2004). *Eendagsvlieg of pionier: welke ondernemer redt onze economie?* Antwerp: Garant.

Maritz, A., & Donovan, J. (2015). Entrepreneurship and innovation: Setting an agenda for greater discipline contextualisation. *Education + Training* 57(1), 74-87.

Schumpeter, J. A. (2010). *Capitalism, socialism and democracy.* London: Routledge.

Robots: friend or foe of the energy-efficient society?

Afman, M., & Scholten, T. (2016). Trends ICT and Energie 2013-2030. Delft. Retrieved from https://www.ce.nl/publicatie/trends_ict_en_energie_2013-2030/1736

Brumson, B. (2018, February 8). Robotics and Energy Cost Reduction. Retrieved from https://www.robotics.org/content-detail.cfm/Industrial-Robotics-Industry-Insights/Robotics-and-Energy-Cost-Reduction/content_id/1047

Duquesne, O. (2018, January 5). Dynamische straatverlichting in Noorwegen. *Autogids*. Retrieved from https://www.autogids.be/autonieuws/video/dynamische-led-straatverlichting-noorwegen.html

Meike, D., Ribickis, L.(2011). Energy Efficient Use of Robotics in the Automobile Industry. In *Proceedings of the 15th IEEE International Conference on Advanced Robotics, Tallinn.*

Will there be a purple bag for robotic waste in the future?

Deahl, D. (2018, April 19). Daisy is Apple's new iPhone-recycling robot. *The Verge*. Retrieved from https://www.theverge.com/2018/4/19/17258180/apple-daisy-iphone-recycling-robot

az, M., et al. (2018), Room-temperature versus heating-mediated healing of a Diels-Alder crosslinked polymer network. *Polymer, 153*, 453–463.

Gsm's and smartphones: tonnen goud? S.d. Geraardpleegd via https://www.recupel.be/nl/waarom-recyclage/gsm-s-en-smartphones-tonnen-goud/

Kumar, V., et al. (Ed.) (2017). *Conducting Hybrid Polymers.* Zwitserland: Springer.

Potenza, A. (2017, November 7). Deep-sea mining could find rare elements for smartphones — but will it destroy rare species? *The Verge.* Retrieved from https://www.theverge.com/2017/10/3/16398518/deep-sea-mining-hydrothermal-vents-japan-precious-metals-rare-species

Scheltjens, G., et al. (2011). Self-healing property characterization of reversible thermoset coatings. *Journal of Thermal Analysis and Calorimetry 105*(3), 805–809.

Terryn, S. et al.(2017). Self-healing soft pneumatic robots. *Science Robotics* 2(9), 1-12.

Torre Muruzabal, A. et al. (2016). Creation of a nanovascular network by electrospun sacrificial nanofibers for self-healing applications and its effect on the flexural properties of the bulk material. *Polymer Testing* 54, 78-83.

Van Damme J. et al.(2017). Anthracene-Based Thiol-Ene Networks with Thermo-Degradable and Photo-Reversible Properties. *Macromolecules* 50(5), 1930-1938.

Vanderborght, B. (2018, April 3). *Toekomst 2.0. Robots die zichzelf herstellen* [Online Video]. Retrieved from https://www.wtnschp.be/wetenschap/technologie/toekomst-2-0-robots-die-zichzelf-herstellen/

Waste Robotics. Retrieved from https://wasterobotic.com/en/

AUTHOR CURRICULUM VITAE

Katrien Beuls holds a PhD in Computer Science (VUB, 2013) and is active within VUB's Artificial Intelligence Lab as a lecturer and researcher. For her doctorate, she built a tutoring system for the Spanish language that uses AI agents to assess the student's current level and to set out a personal learning path. She currently leads the research group on Evolutionary & Hybrid AI within which she develops new techniques that permit more broadly applicable artificial intelligence based upon the principles of evolutionary biology and representing a combination of symbolic and subsymbolic techniques. Her mission is to build AI systems that will be able to truly understand people.

Joost Brancart holds a PhD in Engineering with a specialization in materials science. As a postdoctoral researcher he supervises research projects on improving the functional properties of materials. Sustainability is an important driving force and topics include recycling and the degradation and biodegradation of materials. The self-healing materials he developed during his PhD also have a major impact on the sustainability and lifespan of the applications in which they are used.

Malaika Brengman is Professor of Marketing, Consumer Behaviour and Market Research at the Faculty of Social Sciences and Solvay Business School (VUB) and heads the Marketing & Consumer Behaviour research cluster. Her research focuses on buying behaviour and motivations, both online and offline, and involves examining the future of shopping. Recently, a particular focus has been the impact of new technologies such as augmented and virtual reality and humanoid robots.

Thomas Crispeels is Professor of Technology & Innovation at the Faculty of Social Sciences and the Solvay Business School (VUB). Thomas is head of the Science, Technology, Innovation and Creativity research group. His research focuses on technology transfer and academic entrepreneurship, and more specifically on the question of how research results can flow from university to society.

Sander De Bock focuses his research on the evaluation and optimisation of industrial exoskeletons. He holds a master's degree in Physical Education and Movement Sciences and has a particular interest in technology. He is part of the Human Physiology research group. In addition to testing exoskeletons, his interest lies in the connection between the movements of the human body during industrial work and electrophysiology.

Laurens De Gauquier holds a master's degree in Applied Economics (VUB). He has been working at the VUB as a PhD researcher since 2015 and is part of the Marketing and Consumer Behaviour research cluster within the faculty of Social Sciences and Solvay Business School. His research focuses on the impact of humanoid robots in retail.

Paul De Hert works on privacy & technology, human rights & criminal law. He is a professor at the Faculty of Law and Criminology (VUB) and director of the research groups Fundamental Rights & Constitutionalism (FRC) and Law, Science, Technology & Society (LSTS). He is also attached to the Tilburg Institute for Law and Technology (TILT).

Emma De Keersmaecker is a PhD student at the Faculty of Physical Education and Physiotherapy (VUB). Her research focuses on the use of virtual reality in gait rehabilitation following stroke, specifically looking at the gait pattern and how it is influenced by virtual reality.

Lars De Laet is an Architectural Engineer and a professor at the Faculty of Engineering Sciences (VUB). Together with his colleagues, he develops lightweight structures, biological building materials and computational design methods for sustainable architecture and infrastructure. He does this using the latest 3D modelling techniques and digital fabrication methods. The team designs, calculates and develops innovative constructions such as tents for refugee camps and architectural structures for events. They also investigate the printing of building components from biomaterials with the aid of robots.

Kevin De Pauw is a professor in the Human Physiology research group. He performs research in the field of Human-Centered Robotics in which prosthetic prototypes are evaluated on amputees, and exoskeletons tested in labs and industrial settings. His research focusses on brain activity relating to the muscles during physical effort and the use of this information to control robotic components.

Philippe De Sutter is Head of Unit and responsible for gynaecology and oncology at the University Hospital (VUB). He specialises in gynaecological surgery, laparoscopy and robotic surgery for both benign and oncological indications. In addition, colposcopy and HPV-related disorders are an important area of interest. De Sutter is a guest professor at the Faculty of Medicine of the VUB and also works as a consultant at AZ Sint-Maria in Halle and ASZ in Aalst.

Nico De Witte is a gerontologist, member of Brubotics, lecturer in research methods and techniques at the Faculty of Psychology and Educational Sciences (VUB), and lecturer at the Faculty of Humanities and Welfare of the University College Ghent. In 2002, he helped to set up research into the needs of the elderly, which has grown into an international study known as Belgian Ageing Studies. In 2013, his thesis on vulnerability in elderly people living at home earned him the title of Doctor in Agogic Sciences. He has participated in various research projects, such as the development of age-friendly cities, political participation of older people and new forms of civic participation, new types of care such as inclusive care (INCCA), informal care in residential care centres, vulnerability, and quality of life of elderly people living at home.

Daniel De Wolf is senior lecturer in Criminal Law and Criminal Procedure (VUB) at the Public Law department. In 2009, he obtained his PhD on the subject 'The role of the judge in finding the truth in the criminal procedure'. He is also a lawyer in Brussels.

Ann Dooms is Professor of Mathematics at the VUB, where she leads the research group Digital Mathematics (DIMA). She specialises in the mathematical aspects of *data science*, in particular digital data representation, analysis, communication, security and forensics. In the latter branch she works on the digital analysis of paintings, historical documents, fonts and fake images. Ann was elected as a member of the Jonge Academie (Young Academy) and the IEEE Information Forensics and Security Committee. She was awarded 2nd Prize in Science Communication from the Royal Flemish Academy of Belgium for Science and the Arts (KVAB) in 2014 and the Senior Cera Award in 2018. In 2014 she gave a TEDx lecture in Brussels and in 2016 she appeared in the Belgian TV documentary 'Herontdekking van de Wereld' (Rediscovery of the world) as she journeyed in the footsteps of Alan Turing.

Shirley A. Elprama is a senior researcher at imec-SMIT-VUB. Since 2011 she has been investigating how robots and people work together in the workplace, and how people use robots in their jobs. In her PhD, she focuses on the acceptance of various types of robots (healthcare robot, collaborative robots, exoskeletons) in different user contexts (residential care centres, hospitals, car factories) by different users (physiotherapists, workers, nurses, surgeons).

Katleen Gabriels is a moral philosopher specialising in computer ethics. She studied Germanic languages and moral philosophy at the Catholic University of Leuven, Ghent University and the University of Helsinki. She obtained her PhD at the VUB based upon research into morality in the virtual world. She also worked at the VUB as a postdoctoral researcher and lecturer and then as assistant professor at Eindhoven University of Technology. She is currently a full-time employee at Maastricht University. Katleen is the author of *Onlife*, which was proclaimed 'Book of the year 2016' by independent liberal think tank Liberales, and was awarded the Willy Calewaert 2018-2019 Chair by VUB's Faculty of Engineering Sciences.

Jo Ghillebert holds master's degrees in Physical Education and Exercise Sciences and in Rehabilitation Sciences and Physiotherapy from Ghent University. He is a PhD student in the research group Human Physiology, where his research relates to the evaluation of lower leg prostheses during daily activities, a field in which psychophysiological and biomechanical aspects are important for mapping out the human response.

Marc Goldchstein is a Commercial Engineer at Solvay Business School (VUB). After his studies he performed research on (academic) entrepreneurship at VUB's Centre for Business Economics. He co-founded SoftCore, one of VUB's first start-ups, and was then involved in two other technology start-ups. Since 2004, he has been back at the VUB working as a practical lecturer in entrepreneurship and as a member of the TechTransfer department.

Hens teaches Economics at the VUB. His research is on the uses and impact of globalisation, past and present.

Veerle Hermans is a part-time lecturer in Ergonomics at the Faculty of Psychology and Educational Sciences (VUB). Her research focuses on optimising physical stresses at work and the impact of innovations on employee well-being. She is also responsible for the ergonomics department at IDEWE (the external service for prevention and protection at work).

Rob Heyman is a researcher at imec-SMIT at the VUB. He is responsible for privacy, ethics, trust and security at City of Things in Antwerp. He is also working on privacy and data protection in smart cities, machine learning, social media and online advertising.

Mireille Hildebrandt is a research professor on Interfacing Law and Technology at the VUB. She works with the Law Science Technology and Society studies (LSTS) research group at the Faculty of Law and Criminology. She also holds a part-time chair in the Computer Science Department of the Faculty of Natural Sciences, Mathematics and Computer Science at Radboud University. Her research focuses on the implications of automated decisions, machine learning and *mindless artificial agency* in the democratic constitutional state. In 2018 she received an ERC Advanced Grant from the European Research Council for her project 'Counting as a Human Being in the Era of Computational Law' (see www.cohubicol.com).

Jonathan Holslag teaches International Politics at the VUB. His research focuses on international security, the position of Europe in the world, and Asia. Recent books are *A political history of the world* (2018), *China's coming war with Asia* (2016) and *De kracht van het paradijs* (2015). In addition to his work at the university, Jonathan is a special advisor to the First Vice-President of the European Commission.

An Jacobs holds a PhD in Sociology and is a part-time lecturer in Qualitative Research Methods (VUB). She leads the Smart Health and Work unit within the SMIT research group (imec VUB). As a founding member of Brubotics, she studied current and future human-robot interaction in healthcare and production environments in various projects.

Charlotte Jewell holds a master's degree in Sociology and Anthropology and works as a researcher at the VUB within the SMIT research group in the Smart Health and Work unit. Her research focuses on human-robot interaction in healthcare and on the development of human-robot relationships.

Erika Joos is responsible for Physical Medicine, Rehabilitation and Sports Medicine at the University Hospital of Brussels. After three years of general medicine in Antwerp and a second master's degree in Sports Medicine she trained to become a rehabilitation doctor. She is a member of the physical medicine recognition committee, chair of the accreditation committee and member of the expert committee for Flanders on admission for therapeutic need in sport, recognised by WADA. As president of Europadonna – an association that represents the interests of breast cancer patients – and committed sportsperson she defends the importance of 'mens sana in corpore sano'.

Dimokritos Kavadias is a lecturer at the Political Sciences department (VUB) and teaches subjects such as research methods. He is also director of the Brussels Information, Documentation and Research Centre (BRIO). His research projects focus on political socialisation, political psychology, education, citizenship and Brussels. Partly thanks to his older brothers he has been addicted to science fiction (SF) and fantasy since he was five years old. It is the dilemmas and political implications of SF that put him on the slippery slope of political science.

Eric Kerckhofs is a full professor of neurological rehabilitation and rehabilitation psychology (VUB). He obtained a doctorate in rehabilitation sciences and physiotherapy and a master's degree in clinical psychology from the VUB. He performs research in the field of rehabilitation psychology and is coordinator of the Rehabilitation Research-Neurorehabilitation research group and co-director of the Centre for Neurosciences.

Nina Lefeber is a PhD student at the Faculty of Physical Education and Physiotherapy (VUB). Her research focuses on the use of robotics in gait rehabilitation after stroke, with a particular focus on the exercise intensity during walking in robotic systems.

Dirk Lefeber is a full professor at the Faculty of Engineering Sciences (VUB). He is head of the Robotics research group and leads the multidisciplinary Brubotics research group. This group investigates how the quality of human life can be improved by means of robotics based upon perspectives from various disciplines.

Johan Loeckx is guest professor and lab manager at the Artificial Intelligence Lab (VUB). His research focuses on the intersection between art, education and AI. He was involved in the design of the security system of the Flemish healthcare platform Vitalink and in 2012 became co-founder of the second Freinet school in Brussels. As well as his academic work, he is an active musician/composer in both the popular and classical genres.

Prof. Dr. **Cathy Macharis** leads VUB's interdisciplinary research group MOBI together with Prof. Joeri Van Mierlo. This group performs research into sustainable solutions in the field of mobility and logistics. Cathy teaches sustainable mobility and logistics, operational and logistics management and supply chain management. She is involved in several national, regional and European research projects on sustainable logistics and urban mobility. She chairs the Brussels Mobility Commission. See also mobi.vub.ac.be.

Joachim Mathieu is the coordinator of the Science Outreach Office of VUB. He has been part of the steering committee for RoboCup Junior Belgium since 2011.

Michel Maus is a lecturer in Tax Law at the VUB. He is mainly concerned with the subjects of tax procedures, tax fraud and tax justice. He is also a partner and co-founder of the law firm Bloom Law and a much sought-after speaker at seminars and in the media.

Romain Meeusen is a full professor at the VUB, and chair of the Human Physiology research group. The theme of the research group is Exercise and the Brain in Health & Disease. Since the academic year 2018-2019 he has been vice-rector for internationalisation.

Marc Noppen has worked as an interventional pneumologist in Belgium and abroad for twenty years and has been CEO of the University Hospital of Brussels for twelve. He is a professor of pneumology and of management and policy in healthcare at the VUB, as well as a guest lecturer in health strategy at Vlerick Business School. He is INSEAD certified in governance (IDP-C 14), and a guest lecturer at several foreign universities.

Ann Nowé is a professor at the VUB, where she leads the lab for Artificial Intelligence. After studying mathematics and computer science (UGent), she became assistant to the Faculty of Engineering at the VUB. Her doctorate was on the intersection between Mathematics, Artificial Intelligence and Control Theory. In 1999 she became a lecturer in the Department of Computer Science in the Faculty of Sciences. Her research focuses mainly on self-learning systems, and in particular Reinforcement Learning. She is also investigating how the models can be made more transparent to facilitate a dialogue with domain experts. Her research covers theoretical aspects, experimental results, and a wide range of applications such as robotics, telecommunications, bioinformatics, security, smart grids and smart devices in an Internet of Things context.

Jo Pierson is a full-time professor attached to the Department of Communication Sciences (Faculty of Social Sciences & Solvay Business School, VUB) and also teaches at the University of Amsterdam (Minor Privacy Studies) and Hasselt University (Department of Computer Sciences). He lectures on social and technological aspects of media and ICT use, privacy, digital media marketing and innovation. He also heads the SMIT research centre (Studies on Media, Innovation and Technology) where, as a senior researcher, he is in charge of the research unit that focuses on data, privacy, ethics and literacy. In this role he is attached to the imec.

Lincy Pyl is Civil Engineer Architect and professor at the Faculty of Engineering Sciences (VUB). Her research domains are the static and dynamic analysis of structures with particular consideration for mechanical behaviour under the influence of special loads such as impact, explosions and fatigue. She is interested in the reliability and safety of lightweight structures, sustainability, and digitisation in the construction industry. She also focuses on the characterisation of the mechanical properties of innovative materials such as 3D-printed metals and fibre-reinforced polymer composites for lightweight applications.

Hubert Rahier has been a professor at the Department of Materials and Chemistry (VUB) since 2001. He specialises in the characterisation and development of new materials. As well as self-healing materials, he is active in the development of non-traditional cements, which are based on by-products and must themselves be recyclable. This is his way of helping to realise a circular economy with minimal environmental impact.

Werner Schirmer is a senior researcher in the Department of Sociology (VUB). He was awarded a doctorate from the LMU Munich and holds a Swedish Docent title from Sweden's Uppsala University. At the VUB he teaches Sociological Theories and conducts research into the impact of new digital technologies on society. Other research topics are social robots, AI, social media, virtual reality and augmented reality.

Luc Steels is emeritus professor of the VUB. After studying at MIT under the direction of Marvin Minsky, in 1983 he founded the VUB AI laboratory, which quickly became a leading centre in Europe. Until 2017, Steels taught in the Computer Science department and about forty PhDs in AI have been granted under his direction. His research covers various domains of AI, from knowledge systems and intelligent robots to models for understanding, producing and learning language and the ethical issues around AI.

Eva Swinnen studied Rehabilitation Sciences and Physiotherapy at the VUB. During her doctoral research, she investigated how robotic systems influence the rehabilitation of patients with neurological disorders. She is currently professor of neurological rehabilitation and rehabilitation technology (VUB). Her research focuses on the use of robotics and technology in the rehabilitation of patients with neurological disorders. She is a board member in organisations such as the Belgian association for neurorehabilitation, the Society of Movement Analysis Laboratories in the Low Lands (SMALLL) and the Brussels Human Robotics Research Center (Brubotics), and is scientific coordinator of VUB's Center for Neurosciences (C4N).

Lynn Tytgat is coordinator of weKONEKT.brussels, an initiative in which the VUB and ULB assume their responsibility for the city and community and contribute to the development of a free, connected, resilient and inclusive urban community. In this context, Lynn works closely with the highly regarded art scene in Brussels to strengthen the synergies between science, technology and art.

Guy Van Assche is a professor at the Department of Materials and Chemistry (VUB). He specialises in studying the relationships between the chemical and physical structure of plastics, the processing of these materials from raw materials to finished products, their intermediate and final properties, and how they change with ageing. His work includes self-healing materials, organic photovoltaic solar cells, and the ageing of corrosion-resistant coatings.

Jean Paul Van Bendegem studied Mathematics and Philosophy at Ghent University. He is emeritus professor at the VUB where taught logic and philosophy of science. He was also director of e Centre for Logic and Philosophy of Science. In addition to his academic publications on the philosophy of mathematics and science, he is also author of a number of books for the general public including *De vrolijke atheïst* (the cheerful atheist), *Elke drie seconden* (Every three seconds) and *Verdwaalde stad* (Lost city). You can also occasionally hear him on the radio, see him on television or read his writing in the press.

Greet Van de Perre is a researcher at VUB's robotics group. Her doctoral research focused on the efficient generation of gestures and body movements for social robots and the design of the social robot Elvis. Her current research is in the field of collaborative robotics.

Stephanie van de Sanden holds an MSc in Management (VUB). In April 2016, she joined the VUB within the Marketing and Consumer Behaviour research cluster at the Faculty of Social Sciences and Solvay Business School, where she was appointed to a Baekeland mandate in collaboration with industrial partner Digitopia and sponsored by VLAIO. Her research focuses on the strategic application of innovative digital technologies in physical retail.

Jolien van Keulen studied Film and Television Studies at Utrecht University and Media Studies at the Erasmus University Rotterdam. Since 2014 she has been working as a PhD student and teaching assistant at the Department of Communication Studies (VUB). Her research focuses on TV formats and the transnationalisation of television production.

Jef Van Laer has been working at VUB's Science Outreach office since 2010. He coordinates various scientific communication activities, edits the popular science website wtnschp.be and is a member of the steering group of the Brussels STEM support centre.

Bram Vanderborght is a professor of robotics at the VUB, a member of VUB's robotics research centre Brubotics, and core lab manager of Flanders Make. He is developing a new generation of robots to meet the social challenges of the ageing population and increasing needs in the care sector, including prosthetics and exoskeletons to help people walk. He also devises social robots with emotions and gestures that are used as therapy assistants for children with special needs, and develops cobots for industry. Central to his approach is safe and intuitive collaboration between man and robot.

Prof. **Lieselot Vanhaverbeke** is a senior lecturer at the VUB and teaches in the fields of Operational Research and Research Methods. She is attached to the Business Technology and Operations (BUTO) department and a member of the MOBI (Mobility, Logistics and Automotive Technology) research group. Her research in the field of mobility focuses firstly on socio-economical aspects of electric and autonomous vehicles, such as market developments and the behaviour of mobility users, and secondly on the location-based decisions associated with this field, such as location analysis for charging infrastructure.

Christophe Vanroelen is a senior lecturer at the Department of Sociology (VUB) and chair of the Interface Demography research group. He teaches general sociology, welfare state sociology and labour sociology. His research is mainly in the field of social determinants of health, social inequality in health and work-related health inequality. He is also a member of the Health Inequalities Research Group (GREDS) of Pompeu Fabra University in Barcelona.

Tom Verstraten is a postdoctoral researcher in the robotics research group (VUB) and is supported by a postdoctoral research mandate from the FWO. His research focuses on energy-efficient drives for machines with variable loads and speeds, with special consideration for the integration of elasticity and redundancy.

Lennert Vierendeels is a commercial engineer from the Solvay Business School (VUB). He started his career at Ormit Belgium, but in 2015 his passion for technology brought him back to his alma mater where he is now Business Developer for BruBotics. As a business developer, Lennert is responsible for the economic exploitation of research, including the creation of new robotics spin-offs and the guidance of robotics researchers who want to make the leap into entrepreneurship. Lennert is also treasurer at the Solvay Schools Alumni non-profit organisation.

Kim Willems is a commercial engineer by training and obtained her PhD on the subject of differentiation strategies in retail in 2012 at the VUB and Hasselt University. She is currently Professor of Marketing at the VUB and attached to the Marketing and Consumer Behaviour research cluster, where she researches technology in retailing and services. Her focus is on how an optimal balance between 'tech vs. touch' can improve both customer experiences and retailer performance.